# Principles and Techniques of Molecular Biology

# Principles and Techniques of Molecular Biology

Simon Pade

R CALLISTO REFERENCE

www.callistoreference.com

**Callisto Reference,**
118-35 Queens Blvd., Suite 400,
Forest Hills, NY 11375, USA

Visit us on the World Wide Web at:
www.callistoreference.com

ISBN: 978-1-64116-552-5 (Hardback)

**Cataloging-in-Publication Data**

Principles and techniques of molecular biology / Simon Pade.
    p. cm.
Includes bibliographical references and index.
ISBN 978-1-64116-552-5
1. Molecular biology. 2. Molecular biology--Technique. 3. Biomolecules. 4. Molecules. I. Pade, Simon.
QH506 .P75 2022
572.8--dc23

# Table of Contents

**Permissions**

**Index**

# Preface

The branch of biology that deals with the study of the molecular basis of biological activity between biomolecules in different systems of a cell is referred to as molecular biology. It includes the interactions between DNA, RNA, proteins and their biosynthesis along with the regulation of these interactions. Molecular biology also makes use of techniques and concepts from the fields of genetics and biochemistry. Some of the techniques used within this field are molecular cloning, polymerase chain reaction, gel electrophoresis and allele-specific oligonucleotide. There are numerous processes which are studied within molecular biology such as replication, transcription and cell function. This book is a compilation of chapters that discuss the most vital concepts in the field of molecular biology. While understanding the long-term perspectives of the topics, it makes an effort in highlighting their impact as a modern tool for the growth of the discipline. This textbook will provide comprehensive knowledge to the readers.

Given below is the chapter wise description of the book:

Chapter 1- The domain of biology which deals with the molecular basis of biological activity between biomolecules in different systems of a cell is called molecular biology. The topics elaborated in this chapter will help in gaining a better perspective about molecular biology as well as its principles.

Chapter 2- There are various important concepts within molecular biology. Some of them are cell division, cell cycle, DNA replication, transcription and protein synthesis. This chapter closely examines these key concepts of molecular biology to provide an extensive understanding of the subject.

Chapter 3- Molecular biology makes use of diverse techniques. Some of them are molecular cloning, polymerase chain reaction, gel electrophoresis, blotting and DNA microarray. The diverse applications of these techniques in the field of molecular biology have been thoroughly discussed in this chapter.

Chapter 4- The linear order of DNA elements as well as their division into chromosomes is referred to as genomic organization. DNA sequencing refers to the process which is used to determine the nucleic acid sequence. This chapter has been carefully written to provide an easy understanding of these processes related to the organization and sequencing of genomes and DNA as well as the technologies used for them.

Chapter 5- The process through which information from a gene gets used to synthesize a functional gene product is known as gene expression. The mechanisms which act to repress or induce genes are called gene regulation. This chapter discusses in detail the concepts and processes related to the expression, regulation and translation of genes in eukaryotes and prokaryotes.

Chapter 6- There are a number of different techniques and processes which are involved in the editing or modification of the DNA. A few of them are DNA mutations, DNA repair, genetic recombination, RNA interference and CRISPR. This chapter closely examines the key concepts related to these techniques to provide an extensive understanding of the subject.

At the end, I would like to thank all those who dedicated their time and efforts for the successful completion of this book. I also wish to convey my gratitude towards my friends and family who supported me at every step.

**Simon Pade**

# Chapter 1

## Molecular Biology: An Introduction

The domain of biology which deals with the molecular basis of biological activity between biomolecules in different systems of a cell is called molecular biology. The topics elaborated in this chapter will help in gaining a better perspective about molecular biology as well as its principles.

Molecular biology is the study of living things at the level of the molecules which control them and make them up. While traditional biology concentrated on studying whole living organisms and how they interact within populations (a "top down" approach), molecular biology strives to understand living things by examining the components that make them up (a "bottom up" approach). Both approaches to biology are equally valid, although improvements to technology have permitted scientists to concentrate more on the molecules of life in recent years.

Molecular biology is a specialised branch of biochemistry, the study of the chemistry of molecules which are specifically connected to living processes. Of particular importance to molecular biology are the nucleic acids (DNA and RNA) and the proteins which are constructed using the genetic instructions encoded in those molecules. Other biomolecules, such as carbohydrates and lipids may also be studied for the interactions they have with nucleic acids and proteins. Molecular biology is often separated from the field of cell biology, which concentrates on cellular structures (organelles and the like), molecular pathways within cells and cell life cycles.

The molecules which form the basis of life provide scientists with a more predictable and mechanistic tool for scientists to study. Working with whole organisms (or even just whole cells) can be unpredictable, with the outcome of experiments relying on the interaction of thousands of molecular pathways and external factors. Molecular biology provides scientists with a toolkit with which they may "tinker" with the way life works. They may use them to determine the function of single genes or proteins, and find out what would happen if that gene or protein was absent or faulty. Molecular biology is used to examine when and why certain genes are switched "on" or "off". An understanding of each of the factors has granted scientists a deeper understanding of how living things work, and used this knowledge to develop treatments for when living things don't work so well.

### Common Molecular Biology Techniques

The following list covers some of the more commonly used molecular biology techniques – it is by no means exhaustive:

- Electrophoresis – a process which separates molecules such as DNA or proteins out according to their size, electrophoresis is a mainstay of molecular biology laboratories. While knowing the size of a molecule might not seem like all that much information, it can be used to identify molecules or fragments of molecules and as a check to make sure that we have the correct molecule present.

- Polymerase Chain Reaction (PCR) – a process used to amplify very small amounts of DNA to amounts which can be used in further experiments. It is used as a basic tool in molecular biology to ensure that we have sufficient DNA to carry out further techniques such as genetic modification; however it has wider practical uses such as in forensics (identification using DNA profiling) and disease diagnosis. PCR can also be used to introduce small point mutations into a gene in a process called site-directed mutagenesis.

- Restriction Digest – the process of cutting DNA up into smaller fragments using enzymes which only act at a particular genetic sequence.

- Ligation – the process of joining two pieces of DNA together. Ligation is useful when introducing a new piece of DNA into another genome.

- Blotting – a technique used to specifically identify biomolecules following electrophoresis. The molecule of interest is indicated using either a labeled probe (a complementary strand of nucleic acid) or a labeled antibody raised against a specific protein.

- Cloning – the technique of introducing a new gene into a cell or organism. This can be used to see what effect the expression of that gene has on the organism, to turn the organism into a factory which will produce large quantities of the gene or the protein it codes for, or (within the inclusion of a label) to indicate where the products of that gene are expressed in the organism. Insertion of genetic material into a bacterium is called transformation, while insertion into a eukaryotic cell is called transfection. If a virus is used to introduce this material, the process is called transduction.

Each of these techniques is used in conjunction with other techniques to help scientists solve a particular research question. For example, following using PCR to create large quantities of a particular gene a scientist may ligate a gene for a particular protein into a plasmid vector (a short circular strand of DNA which acts as a carrier), perform a quick restriction digest and electrophoresis to ensure that the gene has been inserted properly, and then use that plasmid to transform a bacterial cell which is used to produce large quantities of the vector. After purification of the vector from the bacteria, it is then used to transfect a mammalian cell in culture. The scientist then uses protein electrophoresis and western blotting to demonstrate the expression of the gene product.

## Principles of Molecular Biology

### DNA

DNA (deoxyribonucleic acid) is the genetic material of cells. It is composed of individual units called nucleotides. A nucleotide is composed of three subunits: a five-carbon sugar (deoxyribose), a phosphate group, and a base. There are four types of bases in DNA: adenine (A), guanine (G), thymine (T), and cytosine (C). Adenine and guanine are purines, and thymine and cytosine are pyrimidines. A polynucleotide chain is formed by linking the adjacent nucleotides via $5' \rightarrow 3'$ phosphodiester bonds. In 1953, Watson and Crick solved the structure of DNA – demonstrating that a DNA molecule is composed of two complementary polynucleotide chains forming the double-helix structure. The double-helix chains are stabilized by hydrogen bonds formed between the oppos-ing A–T and C–G bases on the two complementary polynucleotide chains.

## RNA

In contrast to DNA, RNA (ribonucleic acid) contains the sugar ribose instead of deoxyribose, and uracil (U) instead of thymine (T). An RNA molecule is single-stranded and less stable than a DNA molecule. Cellular RNA serves diverse functions, carried out by different families of RNA mole-cules, including messenger RNA (mRNA), ribosomal RNA (rRNA), transfer RNA (tRNA), small interfering RNA (siRNA), and microRNA (miRNA). Ribozymes are RNA molecules with catalytic 14 J. Zhuge and W. Zhangfunctions. Because RNA molecules have diverse functions, proponents of the "RNA world" theory suggest that RNA may have preceded DNA and protein in life's long evolutionary journey.

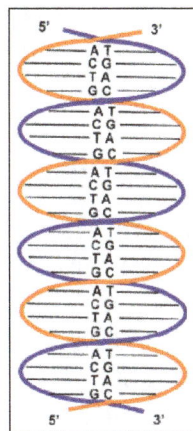

Double-helix structure of DNA: The double-helix structure of DNA is stabilized by the hydrogen bonds formed between A–T and C–G bases on the two antiparallel complementary strands. (A) adenine, (T) thymine, (G) guanine, (C) cytosine.

## Human Genome

The human genome is composed of slightly more than three billion base pairs of DNA, organized into 46 chromosomes (22 pairs of autosomes and 2 sex chromosomes). The Human Genome Proj-ect (HGP) has taught us that there are approximately 20,000–25,000 protein-coding genes, rep-resenting only ~1.5% of the entire genome size. The remainder of the human genome includes regulatory sequences, RNA genes, pseudo genes, repeat sequences, and other sequences for which no known functions are currently understood.

## Repeat DNA Sequences in the Genome

Much of the so-called "junk DNA" (noncoding DNA) in the human genome contains repeat se-quences. There are two types of repeat sequences: (a) tandem repeat and (b) interspersed repeat.

Tandem repeats are two or more nucleotides repeated as a unit one after another in the same orientation. Examples of tandem repeats include microsatellites (2–6 nucleotides long) and mini-satellites (longer than microsatellites, but shorter than 60 nucleotides in length). Microsatellites are useful markers for identity testing, bone marrow chimerism study, microsatellite instability (MSI) testing, and gene dosage studies, such as loss of heterozygosity (LOH) or gene duplication analyses.

Interspersed repetitive DNA sequences, also known as retrotransposons, are repeat elements characterized by RNA intermediates. In mammals, interspersed repetitive DNA constitutes approximately half of the genome. There are two types of interspersed repeat: (a) LTR (long termi-nal repeat) retrotransposons and (b) non-LTR retrotransposons. Non-LTR retrotransposons are fur-ther divided into SINEs (short interspersed nuclear elements) and LINEs (long interspersed nuclear elements). LINEs bear similarities to retroviruses, in that they encode the enzyme reverse-transcriptase which transcribes LINEs RNA into many DNA copies for integration into new genomic loci, thus providing a mechanism for genomic expansion. However, LINEs do not have the LTRs found in retroviruses (i.e., they are not functional retroviruses). LINEs account for approxi-mately 20% of the human genome. SINEs are typically less than 500 bases in length, and do not encode reverse-transcriptase. They constitute about 14% of the human genome. The most common SINEs in the human genome are Alu sequences.

## Eukaryotic Gene Structure and Function

A gene is the hereditary unit of a living organism. The classical concept of genes is centered around the notion that one gene encodes for one protein/enzyme. A classical eukaryotic gene is composed of exons, introns, and regulatory sequences. Exons are stretches of DNA sequences that are represented in the mature form of RNA, including mRNA and tRNA; while, introns are the intervening DNA sequences between exons that will be spliced from the maturing RNA molecule. An RNA transcript usually consists of multiple exons spliced together. A single gene may produce several different transcripts by alternative splicing. The regulatory sequences of a gene include promoters, enhancers, silencers, insulators, and locus control regions (LCRs). A promoter is a region of DNA that facilitates transcription by binding to transcription factors (TFs) and RNA polymerase II. A gene can have several promoters, usually located upstream of the transcription start site. The location of the promoter is designated by counting back from the transcription start site (i.e., −34 refers to 34 base pairs upstream). An enhancer is another type of gene regulatory element that is located either upstream or downstream of the gene, and which regulates gene expression from a greater distance.

Large-scale genomic studies have begun to challenge the classical concept of genes. Data from the International Encyclopedia of DNA Elements (ENCODE) project revealed that genes are surprisingly flexible in the sense that "genes know no borders" (i.e., when a gene is transcribed, the transcript often contains not only the gene itself, but also a portion of the next gene). Such fusion transcripts are estimated to constitute 4–5% of the traditionally recognized gene sequences. In addition, a large number of novel transcription start sites, many of which are located hundreds of thousands of bases away from known start sites, as well as new promoters, have been identified. Surprisingly, nearly a quarter of the newly discovered promoters are located at the end of genes, rather than all at their beginning, as originally thought.

Only 1–2% of human DNA sequences code for proteins. However, genomic studies have shown that much of this noncoding "junk DNA" is transcribed. Among thousands of RNA molecules that are transcribed from the noncoding DNA, the family of functional noncoding RNA (ncRNA) continues to expand. This family now includes tRNA, rRNA, miRNA, siRNA, small nuclear RNA (snRNA), piwi-interacting RNA (piRNA), and long ncRNA.

## Telomere

A chromosome is a thread-like structure composed of a long strand of DNA and associated proteins. The chromosomal ends in eukaryotes are sealed and stabilized by special regions called telomeres. DNA at the telomeric regions is characterized by tandem repeat sequences. For example, human telomeres consist of 2–50 kilobases of TTAGGG tandem repeat sequences.

During DNA replication, the ends of chromosomes cannot be completely replicated, resulting in a shortened copy of DNA. Therefore, telomeric sequences can provide protection against the loss of vital DNA during this process. However, telomeres themselves are subject to shortening during DNA replication, unless they can be replenished by the action of telomerase, a modified RNA polymerase only active in the germ cells of most eukaryotes.

Somatic human cells lack telomerase, and therefore telomeres are shortened during each round of replication in these cells. In addition, oxidative stress can also result in telomere shortening. As telomeres are continuously reduced in length during replication, somatic cells will eventually reach the limit of their replicative capacity and enter senescence. Cellular senescence is thought to play an important role in the suppression of cancer development. The link between telomere and cancer is well established. Cancer cells have found ways to circumvent the replicative limit imposed by shortened telomeres. In fact, most cancer cells possess telomerase activity that can replenish and maintain their telomeres. In addition, some cancer cells may employ an alternative lengthening of telomeres (ALT) pathway, which involves the transfer of telomere tandem repeats between sister chromatids. Telomeres are not only important in cancer research, but also for aging studies. Several premature aging syndromes, such as Werner syndrome, Bloom syndrome, and ataxia-telangiectasia, are characterized by an accelerated rate of telomere attrition. Telomere shortening contributes to stem cell dysfunction and loss of tissue regeneration. However, the use of telomere length or its attrition rate as aging biomarkers in vivo remains to be established.

## Mitochondrial DNA

A mitochondrion contains 2–10 copies of mitochondrial DNA (mtDNA). There are 100–10,000 copies of mtDNA in a human somatic cell. The human mitochondrial genome is a circular DNA molecule with 15,000–17,000 bases, encoding 13 proteins, 22 tRNAs, and one each of the small and large subunits of rRNA. The 13 protein-coding genes of the mitochondrial genome are primarily involved in energy metabolism: subunits 1, 2, and 3 of the cytochrome c oxidase complex; cytochrome b; subunits 6 and 8 of the ATP synthase complex; and six subunits of NADH dehydrogenase. Because human sperms contain far fewer copies of mtDNA than ova, mtDNA typically follows a maternal line of inheritance.

Even though a mitochondrion contains its own DNA, it is nuclear DNA that encodes the majority of its approximate 1,500 proteins, which are transported into the mitochondrion following assembly in the cytoplasm. Therefore, genetic disorders affecting mitochondria can show Mendelian inheritance patterns. Pure mitochondrial genetic disorders show only a maternal pattern of inheritance. Because mitochondria are the "power plants" of the cell, mitochondrial diseases tend to affect organs with high energy requirements, such as muscle, heart, brain, and nerve. Some of the notable mitochondrial diseases include Leber hereditary optic neuropathy (LHON), mitochondrial encephalomyopathy, lactic-acidosis with stroke-like symptoms (MELAS), and myoclonic epilepsy and ragged red fibers (MERRF). Mitochondrial diseases are characterized by considerable heterogeneity, due to variable distribution of defective mtDNA from organ to organ. Of note, frequent mutations in the mitochondrial genome have been reported in both melanoma and non-melanoma skin cancers. Mutant mtDNA in tumor cells might alter mitochondrial-mediated apoptotic pathways to prevent cell death and confer a selective growth advantage.

## Replication, Transcription and Translation

## Replication

Replication of DNA is required to ensure that an exact copy of DNA will be passed down from the maternal cell to its progeny. Watson and Crick first solved the double-helix structure of DNA, and suggested a copying mechanism for DNA replication. Each strand of DNA can serve as a template for the production of a new strand (semiconservative replication). DNA replication in eukaryotes is a parallel process, whereby many chromosomal sites are replicated simultaneously. A new strand of DNA is synthesized in the 5′→3′ direction, because nucleotides can only be added to the 3′ end of the growing nucleotide chain. Replication begins with helicase-mediated unwinding of the double-helix, producing the replication fork and allowing the two existing DNA strands to serve as templates for new strand formation. Only one of the two new strands can be synthesized continuously in the 5′→3′ direction as the replication fork opens. This is called the leading strand. The other strand, which is called the lagging strand, is formed by the joining of many discontinuous small segments (Okazaki fragments) that are synthesized along the lagging strand template. The DNA polymerases involved in lagging and leading strand synthesis also have proof-reading 3′→5′ exonuclease activity.

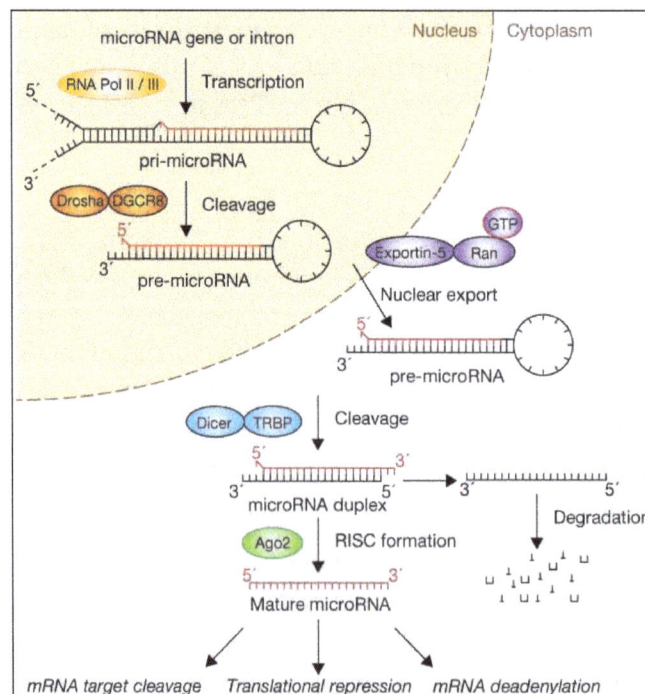

Figure: Biogenesis of miRNA.

Pri-miRNAs are transcribed from miRNA genes by RNA polymerase II or III, under the influence of transcription factors (TF). The pri-miRNAs are processed by Drosha-DGCR8 to pre-miRNAs. In an alternative pathway, miRNAs encoded in the intronic regions ("mirtrons") form pre-miRNAs directly via RNA splicing. The pre-miRNAs are transported from the nucleus to the cytoplasm, and further pro-cessed by Dicer to generate an imperfect double-stranded RNA duplex called miRNA/miRNA. The mature miRNAs contained in RISC (RNA interference silencing complex) bind to specific sites in the 3'-untranslated region of the target mRNA. If base-pairing between the miRNA and its target is perfect, the mRNA will be cleaved. Imperfect pairing between the miRNA and its target can elicit translational repression or mRNA desta-bilization by deadenylation.

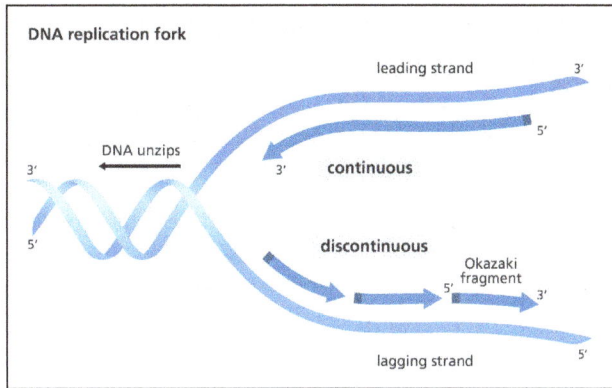

DNA replication.

DNA is unwound by helicase to form the replication fork. The leading strand is synthesized con-tinuously in the 5' to 5' direction. The opposite strand (lagging strand) is formed by joining many discontinuous Okazaki fragments. DNA replication is semiconservative, in that each of the two newly formed DNA copies contains one old strand and one new strand.

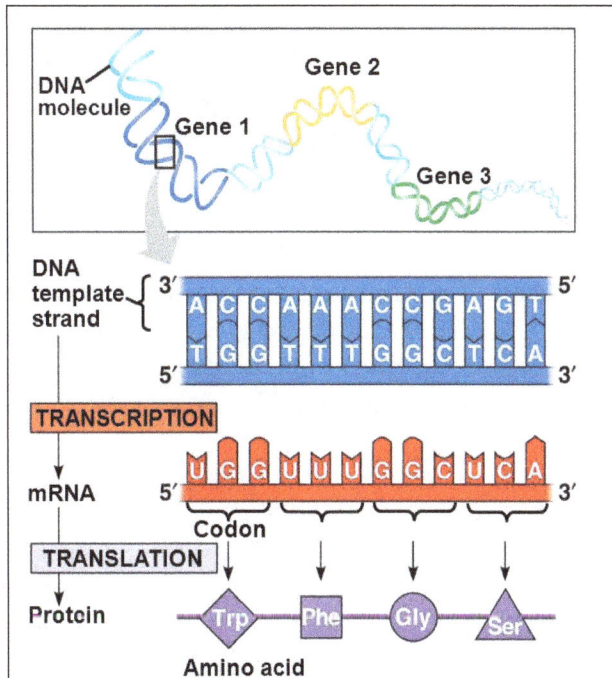

Transcription and translation.

These are two processes that decode genetic information carried by DNA. In transcription, an RNA molecule is synthesized based on its DNA template. The so-called pre-mRNA is processed by addition of a 5' "cap," addition of a 5' polyA tail, and RNA splicing which removes intronic sequences. The mature messenger RNA, which is called mRNA, is then transported to the cytoplasm. In the cytoplasm, the mRNA sequence dictates the synthesis of a polypeptide chain on ribosomes.

## Transcription

The information encoded in DNA dictates RNA synthesis and subsequent protein production. The directional information flow from DNA to RNA, and finally to protein has been called the "central dogma of molecular biology". The transcription of DNA into RNA is a highly regulated process, involving interactions between TFs, promoters and other regulatory elements. Transcription begins at the transcriptional "start site", which lies just upstream of the first coding sequence. From the DNA template, the primary RNA transcript is synthesized in a 5'→3' direction, catalyzed by RNA polymerase II. The primary RNA transcript contains both intron and exon sequences, and is pro-cessed in the nucleus by "capping" at the 5' end and addition of a ploy tail to the 3' end. The RNA transcript is then further processed by removal of its intronic sequences (RNA splicing). The fully processed RNA, now called mRNA, is transported into the cytoplasm where translation takes place.

## Translation

Translation is the process by which a polypeptide chain is synthesized on the basis of the mRNA nucleotide sequence. Translation occurs on ribosomes in the cytoplasm. Eukaryotic ribosomes are composed of large (60S) and small (40S) subunits. The 60S subunit contains 5S, 5.8S, and 28S RNA, and associated proteins. The 40S subunit contains 18S RNA and associated proteins. Translation is mediated by tRNAs, adaptor molecules that have the dual functions of (a) carrying specific amino acids and (b) deciphering the correct codon sequences on mRNA though their anticodon regions. The first translated codon AUG corresponds to the amino acid methionine. The synthesis of a protein involves the successive addition of correct amino acids to the growing polypeptide chain, using mRNA as a template and based on the pairing of the anticodon region of tRNA to a specific codon of mRNA. Translation stops when a stop codon (UAG, UGA, UAA) is reached. A codon is a three-base combination that holds the instructions for translation, either indicating that a particular amino acid should be added or signaling translation initiation or termination. Since there are four different bases (A, T, C, G), the number of possible codons is $4^3$, or 64. However, there are only 20 amino acids. Therefore, more than one codon may encode for one specific amino acid. In such cases, the codon is described as degenerate. Different degenerate codons have identical first two bases, varying only in the third base position.

## Common Types of Mutations

Mutations are changes in DNA sequences. At the nucleotide level, common mutations include point mutations, which can be further defined as silent mutations, nonsense mutations, missense mutations, deletions, and insertions. At the genomic level, mutations include amplifications (gene duplication), interstitial deletions, and chromosomal translocations, or inversions.

## Copy Number Variation

The development and use of new genomic technologies, such as comparative genomic hybridization and microarrays, have resulted in increased recognition of copy number variation (CNV) as a common type of human genetic mutation. Studies of humans from different ethnic backgrounds have shown 1,447 CNV regions, covering ~12% of the human genome. CNVs can involve a single gene or a contiguous set of genes. Variation in the copy number of dosage-sensitive genes may contribute to human phenotypic variability and disease susceptibility.

## Common Types of Mutations

| Type of mutations Characteristics | |
|---|---|
| Nucleotide level | • Silent No change in amino acid |
| | • Missense Change in amino acid |
| | • Nonsense Introduction of a stop codon causing premature termination of translation |
| | • Insertion, deletion Insertion or deletion of nucleotides may result in frameshift |
| Genomic level | • Amplification Multiple copies of a chromosomal region; cause increased gene dosage |
| | • Interstitial deletion Intrachromosomal deletion; may cause gene fusion or loss of heterozygosity |
| | • Translocation Interchange of genetic material from nonhomologous chromosomes |
| | • Inversion Reversing the orientation of a chromosomal segment |
| | • Copy number variation (CNV) Changes in the copy number of a chromosomal segment; can be caused by deletion or duplication. |

## Single Nucleotide Polymorphism

Single nucleotide polymorphism (SNP) refers to a variation in the nucleotide sequence among different individuals of a species. SNP is the most common type of genetic variation, occurring every 100–300 bases in the human genome. The distinction between an SNP and a mutation is rather artificial. In general, if the allele frequency is at least 1%, it is called an SNP; otherwise, it is referred to as a mutation. However, the National Center for Biotechnology Information (NCBI) SNP database (dbSNP) contains SNPs that have allele frequency less than 1%. SNPs can occur in both coding and noncoding regions of the genome. If an SNP is located in the coding region, and it does not change the sequence of the polypeptide chain, it is called synonymous; otherwise, it is termed nonsynonymous.

The study of SNPs will lead to a better understanding of the genotype-phenotype relationship, and help determine an individual's predisposition to common diseases, and their response to the medications used to treat them. For example, studies in methotrexate-treated psoriasis patients suggest that functional SNPs in genes relevant to methotrexate metabolism may influence both the efficacy and toxicity of this drug. In addition, geneticists can use detailed SNP maps and genome-wide association studies (GWAS) to identify disease-causing genes (genetic regions).

## DNA Methylation

DNA methylation is one form of epigenetic regulation, an inheritable influence on gene expression without changes in the DNA sequence. During this process, a methyl group is added to the C5 position of a cytosine pyrimidine ring. In human cells, DNA methylation typically occurs on a cytosine that is followed by a guanine (i.e., CpG dinucleotide). It is estimated that 70% of all CpG sites are methylated in mammals. The unmethylated CpG sites are concentrated in the 5′ upstream region of genes, including the promoter region, forming so-called "CpG islands." Methylation of CpG islands at the promoter region can negatively impact gene expression by blocking the access of TFs. Promoter methylation may play an important role in carcinogenesis. More than half of all known human tumor suppressor genes, including retinoblastoma (RB) and CDKN2A/p16 are subject to promoter methylation in cancer. Promoter methylation status of a select group of genes may serve as biomarkers for disease diagnosis, prognostication, and treatment response prediction. For example, methylation of the MGMT promoter is associated with a favorable response to temozolomide chemotherapy. There is evidence to suggest that epigenetic dysregulation may be associated with not only skin cancers, but also other dermatologic disorders, such as psoriasis, atopic dermatitis, and cutaneous involvement by systemic lupus erythematosus.

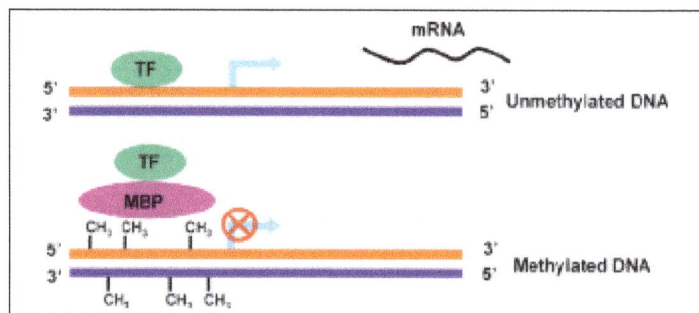

Model of gene expression silencing by promoter methylation.

In the unmethylated state, transcription factor (TF) can bind to the promoter region and activate transcription. If the promoter region is methylated, the methylated CpG site recruits methyl-binding proteins (MBPs), which further recruit other transcription repressors. TF access to the promoter is blocked and transcription is prevented.

J. Zhuge and W. Zhangsubject to promoter methylation in cancer. Promoter methylation status of a select group of genes may serve as biomarkers for disease diagnosis, prognostication, and treatment response prediction. For example, methylation of the MGMT promoter is associated with a favorable response to temozolomide chemotherapy. There is evidence to suggest that epigenetic dysregulation may be associated with not only skin cancers, but also other dermatologic disorders, such as psoriasis, atopic dermatitis, and cutaneous involvement by systemic lupus erythematosus.

# Chapter 2

# Concepts of Molecular Biology

There are various important concepts within molecular biology. Some of them are cell division, cell cycle, DNA replication, transcription and protein synthesis. This chapter closely examines these key concepts of molecular biology to provide an extensive understanding of the subject.

## Cell Division

In unicellular organisms, cell division is the means of reproduction; in multicellular organisms, it is the means of tissue growth and maintenance. Survival of the eukaryotes depends upon interactions between many cell types, and it is essential that a balanced distribution of types be maintained. This is achieved by the highly regulated process of cell proliferation. The growth and division of different cell populations are regulated in different ways, but the basic mechanisms are similar throughout multicellular organisms.

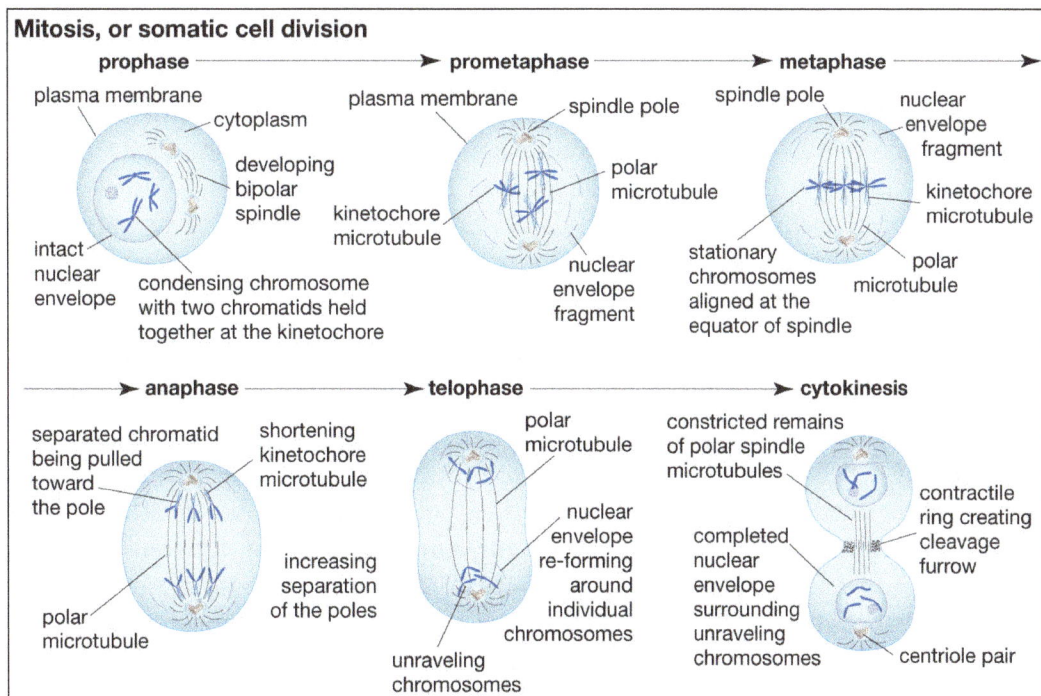

One cell gives rise to two genetically identical daughter cells during the process of mitosis.

Most tissues of the body grow by increasing their cell number, but this growth is highly regulated to maintain a balance between different tissues. In adults most cell division is involved in tissue renewal rather than growth, many types of cells undergoing continuous replacement. Skin cells, for example, are constantly being sloughed off and replaced; in this case, the mature differentiated cells do not divide, but their population is renewed by division of immature stem cells. In certain other cells, such as those of the liver, mature cells remain capable of division to allow growth or regeneration after injury.

In contrast to these patterns, other types of cells either cannot divide or are prevented from dividing by certain molecules produced by nearby cells. As a result, in the adult organism, some tissues have a greatly reduced capacity to renew damaged or diseased cells. Examples of such tissues include heart muscle, nerve cells of the central nervous system, and lens cells in mammals. Maintenance and repair of these cells is limited to replacing intracellular components rather than replacing entire cells.

## Duplication of the Genetic Material

Before a cell can divide, it must accurately and completely duplicate the genetic information encoded in its DNA in order for its progeny cells to function and survive. This is a complex problem because of the great length of DNA molecules. Each human chromosome consists of a long double spiral, or helix, each strand of which consists of more than 100 million nucleotides.

The duplication of DNA is called DNA replication, and it is initiated by complex enzymes called DNA polymerases. These progresses along the molecule, reading the sequences of nucleotides that are linked together to make DNA chains. Each strand of the DNA double helix, therefore, acts as a template specifying the nucleotide structure of a new growing chain. After replication, each of the two daughter DNA double helices consists of one parental DNA strand wound around one newly synthesized DNA strand.

In order for DNA to replicate, the two strands must be unwound from each other. Enzymes called helicases unwind the two DNA strands, and additional proteins bind to the separated strands to stabilize them and prevent them from pairing again. In addition, a remarkable class of enzyme called DNA topoisomerase removes the helical twists by cutting either one or both strands and then resealing the cut. These enzymes can also untangle and unknot DNA when it is tightly coiled into a chromatin fibre.

In the circular DNA of prokaryotes, replication starts at a unique site called the origin of replication and then proceeds in both directions around the molecule until the two processes meet, producing two daughter molecules. In rapidly growing prokaryotes, a second round of replication can start before the first has finished. The situation in eukaryotes is more complicated, as replication moves more slowly than in prokaryotes. At 500 to 5,000 nucleotides per minute (versus 100,000 nucleotides per minute in prokaryotes), it would take a human chromosome about a month to replicate if started at a single site. Actually, replication begins at many sites on the long chromosomes of animals, plants, and fungi. Distances between adjacent initiation sites are not always the same; for example, they are closer in the rapidly dividing embryonic cells of frogs or flies than in adult cells of the same species.

Accurate DNA replication is crucial to ensure that daughter cells have exact copies of the genetic information for synthesizing proteins. Accuracy is achieved by a "proofreading" ability of the DNA polymerase itself. It can erase its own errors and then synthesize a new. There are also repair systems that correct genetic damage to DNA. For example, the incorporation of an incorrect nucleotide, or damage caused by mutagenic agents, can be corrected by cutting out a section of the daughter strand and recopying the parental strand.

## Mitosis and Cytokinesis

In eukaryotes the processes of DNA replication and cell division occur at different times of the cell division cycle. During cell division, DNA condenses to form short, tightly coiled, rod like chromosomes. Each chromosome then splits longitudinally, forming two identical chromatids. Each pair of chromatids is divided between the two daughter cells during mitosis, or division of the nucleus, a process in which the chromosomes are propelled by attachment to a bundle of microtubules called the mitotic spindle.

Mitosis can be divided into five phases. In prophase the mitotic spindle forms and the chromosomes condense. In prometaphase the nuclear envelope breaks down (in many but not all eukaryotes) and the chromosomes attach to the mitotic spindle. Both chromatids of each chromosome attach to the spindle at a specialized chromosomal region called the kinetochore. In metaphase the condensed chromosomes align in a plane across the equator of the mitotic spindle. Anaphase follows as the separated chromatids move abruptly toward opposite spindle poles. Finally, in telophase a new nuclear envelope forms around each set of unraveling chromatids.

An essential feature of mitosis is the attachment of the chromatids to opposite poles of the mitotic spindle. This ensures that each of the daughter cells will receive a complete set of chromosomes. The mitotic spindle is composed of microtubules, each of which is a tubular assembly of molecules of the protein tubulin. Some microtubules extend from one spindle pole to the other, while a second class extends from one spindle pole to a chromatid. Microtubules can grow or shrink by the addition or removal of tubulin molecules. The shortening of spindle microtubules at anaphase propels attached chromatids to the spindle poles, where they unravel to form new nuclei.

The two poles of the mitotic spindle are occupied by centrosomes, which organize the microtubule arrays. In animal cells each centrosome contains a pair of cylindrical centrioles, which are themselves composed of complex arrays of microtubules. Centrioles duplicate at a precise time in the cell division cycle, usually close to the start of DNA replication.

After mitosis comes cytokinesis, the division of the cytoplasm. This is another process in which animal and plant cells differ. In animal cells cytokinesis is achieved through the constriction of the cell by a ring of contractile microfilaments consisting of actin and myosin, the proteins involved in muscle contraction and other forms of cell movement. In plant cells the cytoplasm is divided by the formation of a new cell wall, called the cell plate, between the two daughter cells. The cell plate arises from small Golgi-derived vesicles that coalesce in a plane across the equator of the late telophase spindle to form a disk-shaped structure. In this process, each vesicle contributes its membrane to the forming cell membranes and its matrix contents to the forming cell wall. A second set of vesicles extends the edge of the cell plate until it reaches and fuses with the sides of the parent cell, thereby completely separating the two new daughter cells. At this point, cellulose synthesis commences, and the cell plate becomes a primary cell wall.

## Meiosis

A specialized division of chromosomes called meiosis occurs during the formation of the reproductive cells, or gametes, of sexually reproducing organisms. Gametes such as ova, sperm, and pollen begin as germ cells, which, like other types of cells, have two copies of each gene in their nuclei. The chromosomes composed of these matching genes are called homologs. During DNA replication, each chromosome duplicates into two attached chromatids. The homologous chromosomes are then separated to opposite poles of the meiotic spindle by microtubules similar to those of the mitotic spindle. At this stage in the meiosis of germ cells, there is a crucial difference from the mitosis of other cells. In meiosis the two chromatids making up each chromosome remain together, so that whole chromosomes are separated from their homologous partners. Cell division then occurs, followed by a second division that resembles mitosis more closely in that it separates the two chromatids of each remaining chromosome. In this way, when meiosis is complete, each mature gamete receives only one copy of each gene instead of the two copies present in other cells.

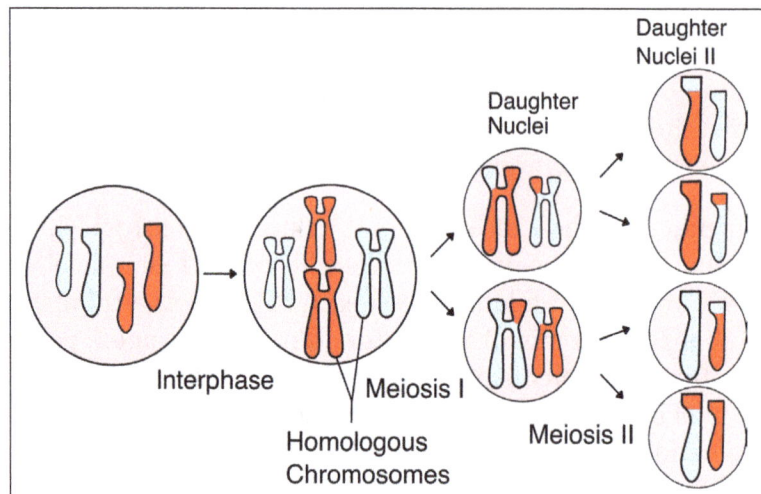

The formation of gametes (sex cells) occurs during the process of meiosis.

# Cell Cycle

The most basic function of the cell cycle is to duplicate accurately the vast amount of DNA in the chromosomes and then segregate the copies precisely into two genetically identical daughter cells. These processes define the two major phases of the cell cycle. DNA duplication occurs during S phase (S for synthesis), which requires 10–12 hours and occupies about half of the cell-cycle time in a typical mammalian cell. After S phase, chromosome segregation and cell division occur in M phase (M for mitosis), which requires much less time (less than an hour in a mammalian cell). M phase involves a series of dramatic events that begin with nuclear division, or mitosis. Mitosis begins with chromosome condensation: the duplicated DNA strands, packaged into elongated chromosomes, condense into the much more compact chromosomes required for their segregation. The nuclear envelope then breaks down, and the replicated chromosomes, each consisting of a pair of sister chromatids, become attached to the microtubules of the mitotic spindle. As mitosis proceeds, the cell pauses briefly in a state called metaphase, when the chromosomes are aligned at the equator of the mitotic spindle, poised for segregation. The sudden separation of sister chromatids marks the beginning of anaphase, during which the chromosomes move to opposite poles of the spindle, where they decondense and reform intact nuclei. The cell is then pinched in two by cytoplasmic division, or cytokinesis, and cell division is complete.

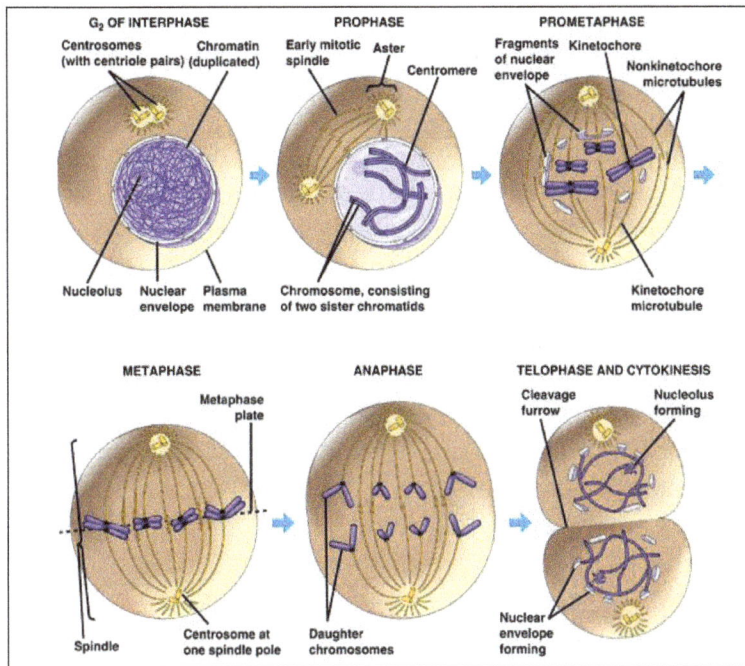

Figure: The events of eucaryotic cell division.

The easily visible processes of nuclear division (mitosis) and cell division (cytokinesis), collectively called M phase, typically occupy only a small fraction of the cell cycle. The other, much longer, part of the cycle is known as interphase. The five stages of mitosis are shown: an abrupt change in the biochemical state of the cell occurs at the transition from metaphase to anaphase. A cell can pause in metaphase before this transition point, but once the point has been passed, the cell carries on to the end of mitosis and through cytokinesis into interphase. DNA replication occurs in interphase. The part of interphase where DNA is replicated is called S phase.

Most cells require much more time to grow and double their mass of proteins and organelles than they require to replicate their DNA and divide. Partly to allow more time for growth, extra gap phases are inserted in most cell cycles—a $G_1$ phase between M phase and S phase and a $G_2$ phase between S phase and mitosis. Thus, the eucaryotic cell cycle is traditionally divided into four sequential phases: $G_1$, S, $G_2$, and M. $G_1$, S, and $G_2$ together are called interphase. In a typical human cell proliferating in culture, interphase might occupy 23 hours of a 24 hour cycle, with 1 hour for M phase.

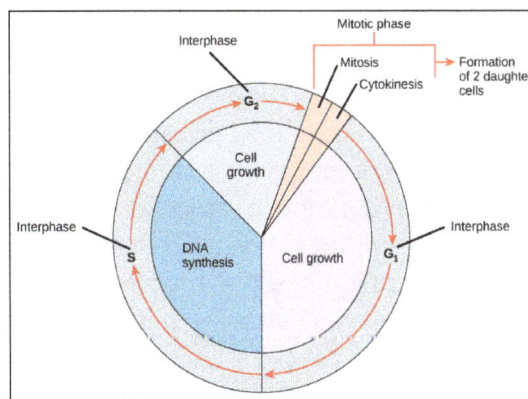

Figure: The phases of the cell cycle.

The cell grows continuously in interphase, which consists of three phases: DNA replication is confined to S phase; $G_1$ is the gap between M phase and S phase, while $G_2$ is the gap between S phase and M phase. In M phase, the nucleus and then the cytoplasm divide.

The two gap phases serve as more than simple time delays to allow cell growth. They also provide time for the cell to monitor the internal and external environment to ensure that conditions are suitable and preparations are complete before the cell commits itself to the major upheavals of S phase and mitosis. The $G_1$ phase is especially important in this respect. Its length can vary greatly depending on external conditions and extracellular signals from other cells. If extracellular conditions are unfavorable, for example, cells delay progress through $G_1$ and may even enter a specialized resting state known as $G_0$ (G zero), in which they can remain for days, weeks, or even years before resuming proliferation. Indeed, many cells remain permanently in $G_0$ until they or the organism dies. If extracellular conditions are favorable and signals to grow and divide are present, cells in early $G_1$ or $G_0$ progress through a commitment point near the end of $G_1$ known as Start (in yeasts) or the restriction point (in mammalian cells). After passing this point, cells are committed to DNA replication, even if the extracellular signals that stimulate cell growth and division are removed.

## Cell Cycle Control System in Eucaryotes

Some features of the cell cycle, including the time required to complete certain events, vary greatly from one cell type to another, even in the same organism. The basic organization of the cycle and its control system, however, are essentially the same in all eucaryotic cells. The proteins of the control system first appeared over a billion years ago. Remarkably, they have been so well conserved over the course of evolution that many of them function perfectly when transferred from a human cell to a yeast cell. We can therefore study the cell cycle and its regulation in a variety of organisms and use the findings from all of them to assemble a unified picture of how eucaryotic cells divide.

## Cell Cycle Control System in Yeasts

Yeasts are tiny, single-celled fungi whose mechanisms of cell-cycle control are remarkably similar to our own. Two species are generally used in studies of the cell cycle. The fission yeast Schizosaccharomyces pombe is named after the African beer it is used to produce. It is a rod-shaped cell that grows by elongation at its ends. Division occurs by the formation of a septum, or cell plate, in the center of the rod. The budding yeast Saccharomycescerevisiae is used by brewers, as well as by bakers. It is an oval cell that divides by forming a bud, which first appears during $G_1$ and grows steadily until it separates from the mother cell after mitosis.

(A) The fission yeast has a typical eucaryotic cell cycle with $G_1$, S, $G_2$, and M phases. In contrast with what happens in higher eucaryotic cells, however, the nuclear envelope of the yeast cell does not break down during M phase. The microtubules of the mitotic spindle (light green) form inside the nucleus and are attached to spindle pole bodies (dark green) at its periphery. The cell divides by forming a partition (known as the cell plate) and splitting in two. The condensed mitotic chromosomes (red) are readily visible in fission yeast, but are less easily seen in budding yeasts. (B) The budding yeast has normal $G_1$ and S phases but does not have a normal $G_2$ phase. Instead, a microtubule-based spindle begins to form inside the nucleus early in the cycle, during S phase. In contrast with a fission yeast cell, the cell divides by budding. As in fission yeasts, but in contrast with higher eucaryotic cells, the nuclear envelope remains intact during mitosis, and the spindle forms within the nucleus.

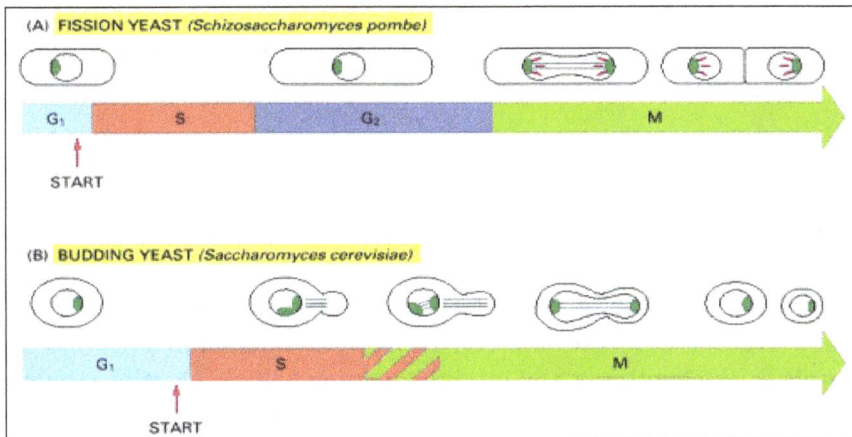

Figure: A comparison of the cell cycles of fission yeasts and budding yeasts.

Despite their outward differences, the two yeast species share a number of features that are extremely useful for genetic studies. They reproduce almost as rapidly as bacteria and have a genome size less than 1% that of a mammal. They are amenable to rapid molecular genetic manipulation, whereby genes can be deleted, replaced, or altered. Most importantly, they have the unusual ability to proliferate in a haploid state, in which only a single copy of each gene is present in the cell. When cells are haploid, it is easy to isolate and study mutations that inactivate a gene, as one avoids the complication of having a second copy of the gene in the cell.

Many important discoveries about cell-cycle control have come from systematic searches for mutations in yeasts that inactivate genes encoding essential components of the cell-cycle control system. The genes affected by these mutations are known as cell-division-cycle genes, or cdc genes. Many of these mutations cause cells to arrest at a specific point in the cell cycle, suggesting that the normal gene product is required to get the cell past this point.

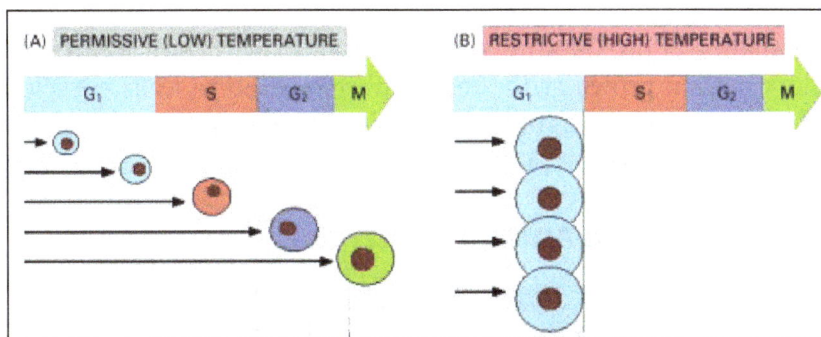

Figure: The behavior of a temperature-sensitive cdc mutant.

A mutant that cannot complete the cell cycle cannot be propagated. Thus, cdc mutants can be selected and maintained only if their phenotype is conditional—that is, if the gene product fails to function only in certain specific conditions. Most conditional cell-cycle mutations are temperature-sensitive mutations, in which the mutant protein fails to function at high temperatures but functions well enough to allow cell division at low temperatures. A temperature-sensitive cdc mutant can be propagated at a low temperature (the permissive condition) and then raised to a higher temperature (the restrictive condition) to switch off the function of the mutant gene. At the higher temperature, the cells continue through the cell cycle until they reach the point where the function of the mutant gene is required for further progress, and at this point they halt. In budding yeasts, a uniform cell-cycle arrest of this type can be detected by just looking at the cells: the presence or absence of a bud, and bud size, indicate the point in the cycle at which the mutant is arrested.

(A) At the permissive (low) temperature, the cells divide normally and are found in all phases of the cycle (the phase of the cell is indicated by its color). (B) On warming to the restrictive (high) temperature, at which the mutant gene product functions abnormally, the mutant cells continue to progress through the cycle until they come to the specific step that they are unable to complete (initiation of S phase, in this example). Because the cdc mutants still continue to grow, they become abnormally large. By contrast, non-cdc mutants, if deficient in a process that is necessary throughout the cycle for biosynthesis and growth (such as ATP production), halt haphazardly at any stage of the cycle—depending on when their biochemical reserves run out.

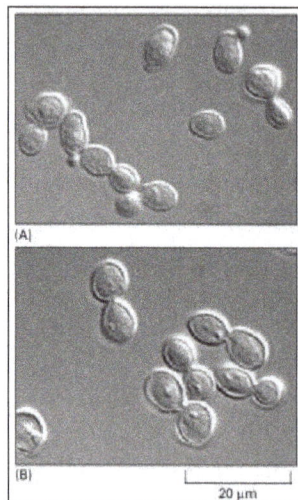

Figure: The morphology of budding yeast cells arrested by a cdc mutation.

(A) In a normal population of proliferating yeast cells, buds vary in size according to the cell-cycle stage. (B) In a cdc15 mutant grown at the restrictive temperature, cells complete anaphase but cannot complete the exit from mitosisand cytokinesis. As a result, they arrest uniformly with the large buds, which are characteristic of late M phase.

## Cell Cycle Control System in Animal Embryos

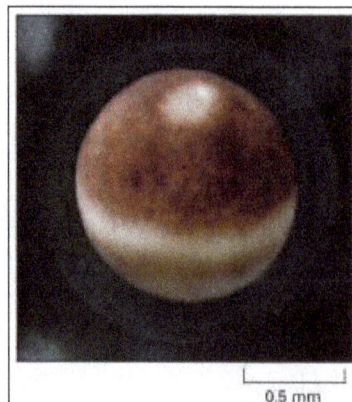

Figure: A mature Xenopus egg, ready for fertilization.

While yeasts are ideal for studying the genetics of the cell cycle, the biochemistry of the cycle is most easily analyzed in the giant fertilized eggs of many animals, which carry large stockpiles of the proteins needed for cell division. The egg of the frog Xenopus, for example, is over 1 mm in diameter and carries 100,000 times more cytoplasm than an average cell in the human body. Fertilization of the Xenopus egg triggers an astonishingly rapid sequence of cell divisions, called cleavage divisions, in which the single giant cell divides, without growing, to generate an embryo containing thousands of smaller cells. In this process, almost the only macromolecules synthesized are DNA—required to produce the thousands of new nuclei and a small amount of protein. After a first division that takes about 90 minutes, the next 11 divisions occur, more or less synchronously, at 30-minute intervals, producing about 4096 ($2^{12}$) cells within 7 hours. Each cycle is divided into S and M phases of about 15 minutes each, without detectable $G_1$ or $G_2$ phases.

The pale spot near the top shows the site of the nucleus, which has displaced the brown pigment in the surface layer of the egg cytoplasm. Although this cannot be seen in the picture, the nuclear envelope has broken down during the process of egg maturation.

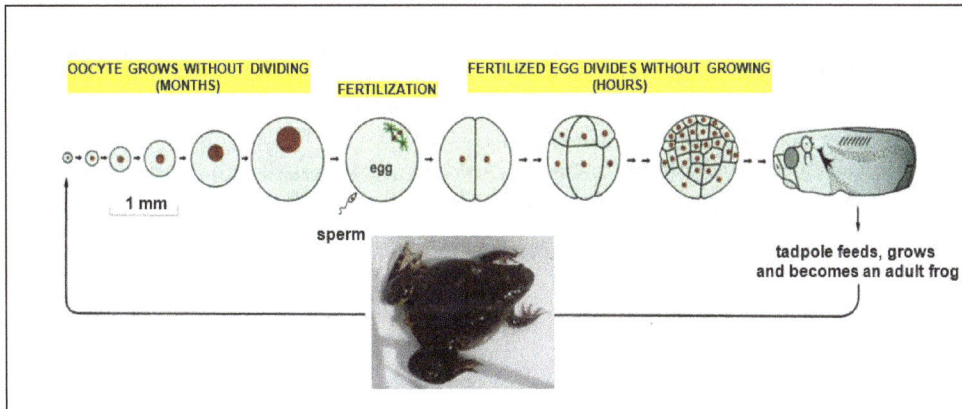

Figure: Oocyte growth and egg cleavage in Xenopus.

The oocyte grows without dividing for many months in the ovary of the mother frog and finally matures into an egg. Upon fertilization, the egg cleaves very rapidly—initially at a rate of one division cycle every 30 minutes—forming a multicellular tadpole within a day or two. The cells get progressively smaller with each division, and the embryo remains the same size. Growth starts only when the tadpole begins feeding. The drawings in the top row are all on the same scale.

The cells in early embryos of Xenopus, as well as those of the clam Spisula and the fruit fly Drosophila, are thus capable of exceedingly rapid division in the absence of either growth or many of the control mechanisms that operate in more complex cell cycles. These early embryonic cell cycles therefore reveal the workings of the cell-cycle control system stripped down and simplified to the minimum needed to achieve the most fundamental requirements—the duplication of the genome and its segregation into two daughter cells. Another advantage of these early embryos for cell-cycle analysis is their large size. It is relatively easy to inject test substances into an egg to determine their effect on cell-cycle progression. It is also possible to prepare almost pure cytoplasm from Xenopus eggs and reconstitute many events of the cell cycle in a test tube. In such cell extracts, one can observe and manipulate cell-cycle events under highly simplified and controllable conditions.

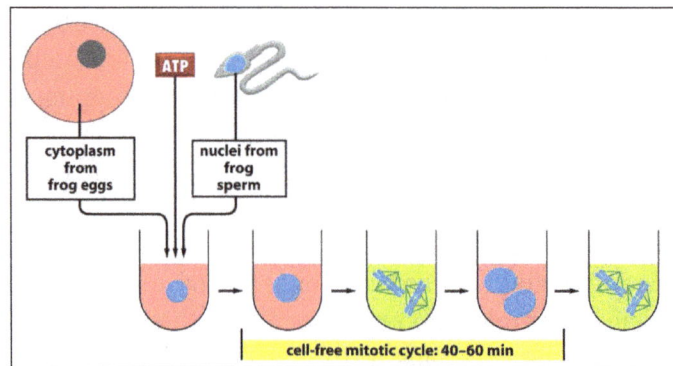

Figure: Studying the cell cycle in a cell-free system.

A large batch of activated frog eggs is broken open by gentle centrifugation, which also separates the cytoplasm from other cell components. The undiluted cytoplasm is collected, and sperm nuclei are added to it, together with ATP. The sperm nuclei decondense and then go through repeated cycles of DNA replication and mitosis, indicating that the cell-cycle control system is operating in this cell-free cytoplasmic extract.

## Cell Cycle Control System of Mammals

It is not easy to observe individual cells in an intact mammal. Most studies on mammalian cell-cycle control therefore use cells that have been isolated from normal tissues or tumors and grown in plastic culture dishes in the presence of essential nutrients and other factors. There is a complication, however. When cells from normal mammalian tissues are cultured in standard conditions, they often stop dividing after a limited number of division cycles. Human fibroblasts, for example, permanently cease dividing after 25–40 divisions, a process called replicative cell senescence. The cells in this scanning electron micrograph are rat fibroblasts.

Mammalian cells occasionally undergo mutations that allow them to proliferate readily and indefinitely in culture as "immortalized" cell lines. Although they are not normal, such cell lines are used widely for cell-cycle studies, and for cell biology generally, because they provide an unlimited source of genetically homogeneous cells. In addition, these cells are sufficiently large to allow detailed cytological observations of cell-cycle events, and they are amenable to biochemical analysis of the proteins involved in cell-cycle control.

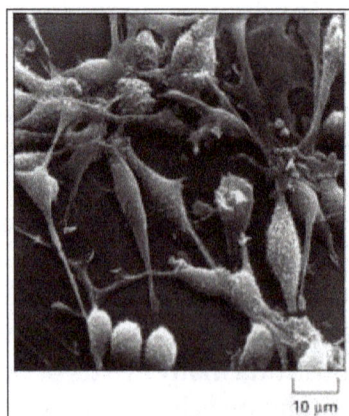

Figure: Mammalian cells proliferating in culture.

Studies of cultured mammalian cells have been especially useful for examining the molecular mechanisms governing the control of cell proliferation in multicellular organisms. Such studies are important not only for understanding the normal controls of cell numbers in tissues but also for understanding the loss of these controls in cancer.

Cell Cycle Progression Can Be Studied in Various Ways. How can one tell at what stage an animal cell is in the cell cycle? One way is to simply look at living cells with a microscope. A glance at a population of mammalian cells proliferating in culture reveals that a fraction of the cells have rounded up and are in mitosis. Others can be observed in the process of cytokinesis. The S-phase cells, however, cannot be detected by simple observation. They can be recognized, however, by supplying them with visualizable molecules that are incorporated into newly synthesized DNA, such as $^3$H-thymidine or the artificial thymidine analog bromo-deoxyuridine (BrdU). Cell nuclei that have incorporated $^3$H-thymidine are visualized by autoradiography, whereas those that have incorporated BrdU are visualized by staining with anti-BrdU antibodies.

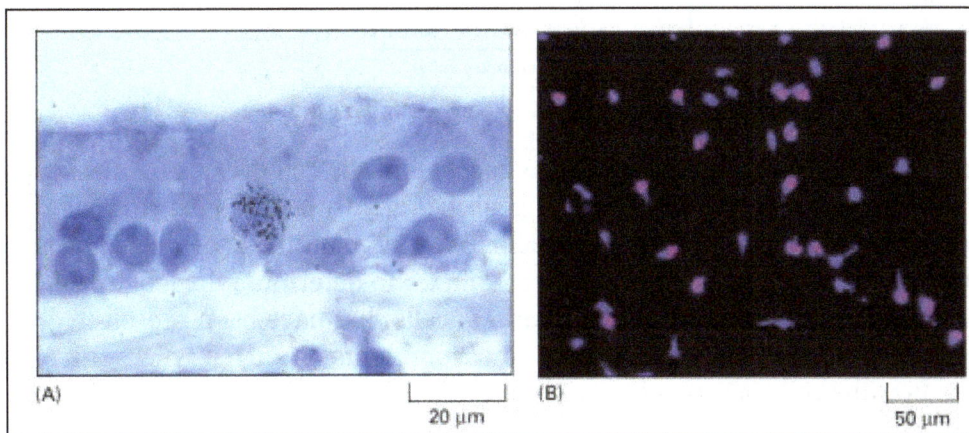

Figure: Labeling S-phase cells.

(A) The tissue has been exposed for a short period to $^3$H-thymidine and the labeled cells have been visualized by autoradiography. Silver grains (black dots) in the photographic emulsion over a nucleus indicate that the cell incorporated $^3$H-thymidine into its DNA and thus was in S phase some time during the labeling period. In this specimen, showing the sensory epithelium from the inner ear of a chicken, the presence of an S-phase cell is evidence of cell proliferation occurring in response to damage. (B) An immunofluorescence micrograph of BrdU-labeled glial precursor cells in culture. The cells were exposed to BrdU for 4 h and were then fixed and labeled with fluorescent anti-BrdU antibodies (red). All the cells are stained with a blue fluorescent dye.

Typically, in a population of cells that are all proliferating rapidly but asynchronously, about 30–40% will be in S phase at any instant and become labeled by a brief pulse of $^3$H-thymidine or BrdU. From the proportion of cells in such a population that are labeled (the labeling index), one can estimate the duration of S phase as a fraction of the whole cell cycle duration. Similarly, from the proportion of these cells in mitosis (the mitotic index), one can estimate the duration of M phase. In addition, by giving a pulse of $^3$H-thymidine or BrdU and allowing the cells to continue around the cycle for measured lengths of time, one can determine how long it takes for an S-phase cell to progress through $G_2$ into M phase, through M phase into $G_1$, and finally through $G_1$ back into S phase.

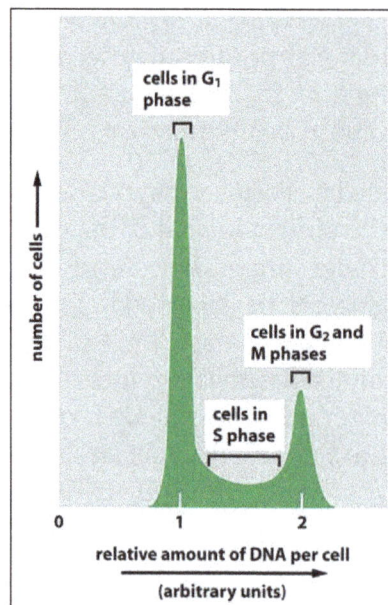

Figure: Analysis of DNA content with a flow cytometer.

Another way to assess the stage that a cell has reached in the cell cycle is by measuring its DNA content, which doubles during S phase. This approach is greatly facilitated by the use of DNA-binding fluorescent dyes and a flow cytometer, which allows large numbers of cells to be analyzed rapidly and automatically. One can also use flow cytometer to determine the lengths of $G_1$, S, and $G_2$ + M phases, by following over time a population of cells that have been preselected to be in one particular phase of the cell cycle: DNA content measurements on such a synchronized population of cells reveal how the cells progress through the cycle.

This graph shows typical results obtained for a proliferating cell population when the DNA content of its individual cells is determined in a flow cytometer. (A flow cytometer, also called a fluorescence-activated cell sorter, or FACS, can also be used to sort cells according to their fluorescence). The cells analyzed here were stained with a dye that becomes fluorescent when it binds to DNA, so that the amount of fluorescence is directly proportional to the amount of DNA in each cell. The cells fall into three categories: those that have an unreplicated complement of DNA and are therefore in $G_1$ phase, those that have a fully replicated complement of DNA (twice the $G_1$ DNA content) and are in $G_2$ or M phase, and those that have an intermediate amount of DNA and are in S phase. The distribution of cells in the case illustrated indicates that there are greater numbers of cells in $G_1$ phase than in $G_2$ + M phase, showing that $G_1$ is longer than $G_2$ + M in this population.

Cell reproduction begins with duplication of the cell's contents, followed by distribution of those contents into two daughter cells. Chromosome duplication occurs during S phase of the cell cycle, whereas most other cell components are duplicated continuously throughout the cycle. During M phase, the replicated chromosomes are segregated into individual nuclei (mitosis), and the cell then splits in two (cytokinesis). S phase and M phase are usually separated by gap phases called $G_1$ and $G_2$, when cell-cycle progression can be regulated by various intracellular and extracellular signals. Cell-cycle organization and control have been highly conserved during evolution, and studies in a wide range of systems—including yeasts, frog embryos, and mammalian cells in culture—have led to a unified view of eucaryotic cell-cycle control.

# DNA Replication

DNA replication, or the copying of a cell's DNA, is no simple task. There are about 3 billion base pairs of DNA in your genome, all of which must be accurately copied when any one of your trillions of cells divides.

The basic mechanisms of DNA replication are similar across organisms. In this topic, we'll focus on DNA replication as it takes place in the bacterium *E. coli*, but the mechanisms of replication are similar in humans and other eukaryotes.

Let's take a look at the proteins and enzymes that carry out replication, seeing how they work together to ensure accurate and complete replication of DNA.

## Basic Idea

DNA replication is semiconservative, meaning that each strand in the DNA double helix acts as a template for the synthesis of a new, complementary strand.

This process takes us from one starting molecule to two "daughter" molecules, with each newly formed double helix containing one new and one old strand.

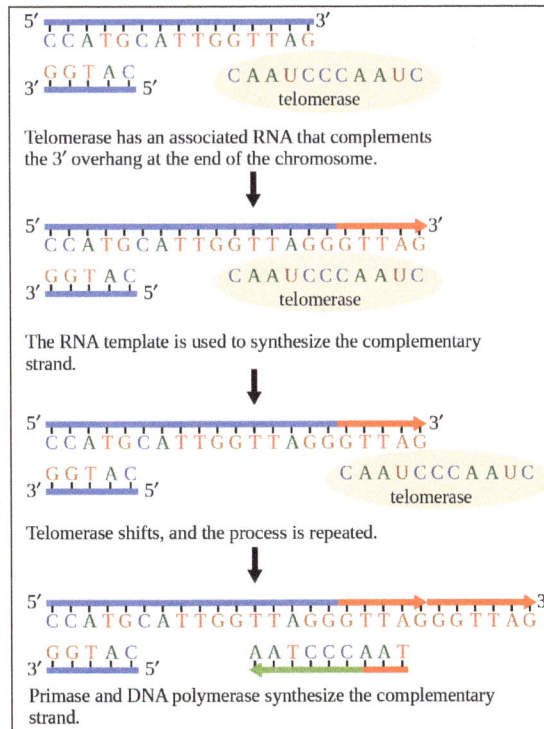

In a sense, that's all there is to DNA replication! But what's actually most interesting about this process is how it's carried out in a cell.

Cells need to copy their DNA very quickly, and with very few errors (or risk problem such as cancer). To do so, they use a variety of enzymes and proteins, which work together to make sure DNA replication is performed smoothly and accurately.

## DNA Polymerase

One of the key molecules in DNA replication is the enzyme DNA polymerase. DNA polymerases are responsible for synthesizing DNA: they add nucleotides one by one to the growing DNA chain, incorporating only those that are complementary to the template.

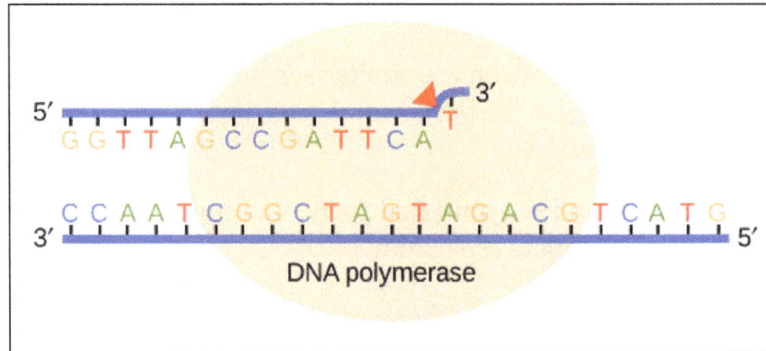

Here are some key features of DNA polymerases:

- They always need a template;

- They can only add nucleotides to the 3′ end of a DNA strand;

- They can't start making a DNA chain from scratch, but require a pre-existing chain or short stretch of nucleotides called a primer;

- They proofread, or check their work, removing the vast majority of "wrong" nucleotides that are accidentally added to the chain.

The addition of nucleotides requires energy. This energy comes from the nucleotides themselves, which have three phosphates attached to them (much like the energy-carrying molecule ATP). When the bond between phosphates is broken, the energy released is used to form a bond between the incoming nucleotide and the growing chain.

In prokaryotes such as E. coli, there are two main DNA polymerases involved in DNA replication: DNA pol III (the major DNA-maker), and DNA pol I, which plays a crucial supporting role we'll examine later.

## Starting DNA Replication

Replication always starts at specific locations on the DNA, which are called origins of replication and are recognized by their sequence.

*E. coli*, like most bacteria, has a single origin of replication on its chromosome. The origin is about 245 base pairs long and has mostly A/T base pairs (which are held together by fewer hydrogen bonds than G/C base pairs), making the DNA strands easier to separate.

Specialized proteins recognize the origin, bind to this site, and open up the DNA. As the DNA opens, two Y-shaped structures called replication forks are formed, together making up what's called a replication bubble. The replication forks will move in opposite directions as replication proceeds.

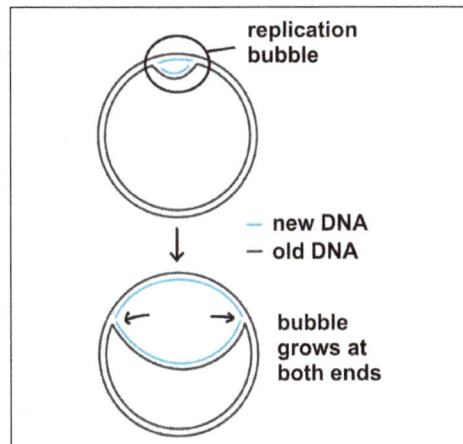

How does replication actually get going at the forks? Helicase is the first replication enzyme to load on at the origin of replication 3' end. Helicase's job is to move the replication forks forward by "unwinding" the DNA (breaking the hydrogen bonds between the nitrogenous base pairs).

Proteins called single-strand binding proteins coat the separated strands of DNA near the replication fork, keeping them from coming back together into a double helix.

## Primers and Primase

DNA polymerases can only add nucleotides to the 3' end of an existing DNA strand. (They use the free -OH group found at the 3' end as a "hook," adding a nucleotide to this group in the polymerization reaction). How, then, does DNA polymerase add the first nucleotide at a new replication fork?

Alone, it can't. The problem is solved with the help of an enzyme called primase. Primase makes an RNA primer or short stretch of nucleic acid complementary to the template that provides a 3' end for DNA polymerase to work on. A typical primer is about five to ten nucleotides long. The primer primes DNA synthesis, i.e., gets it started.

Once the RNA primer is in place, DNA polymerase "extends" it, adding nucleotides one by one to make a new DNA strand that's complementary to the template strand.

## Leading and Lagging Strands

In *E. coli*, the DNA polymerase that handles most of the synthesis is DNA polymerase III. There are two molecules of DNA polymerase III at a replication fork, each of them hard at work on one of the two new DNA strands.

DNA polymerases can only make DNA in the 5' to 3' direction, and this poses a problem during replication. A DNA double helix is always anti-parallel; in other words, one strand runs in the 5' to 3' direction, while the other runs in the 3' to 5' direction. This makes it necessary for the two new strands, which are also antiparallel to their templates, to be made in slightly different ways.

One new strand, which runs 5' to 3' towards the replication fork, is the easy one. This strand is made continuously, because the DNA polymerase is moving in the same direction as the replication fork. This continuously synthesized strand is called the leading strand.

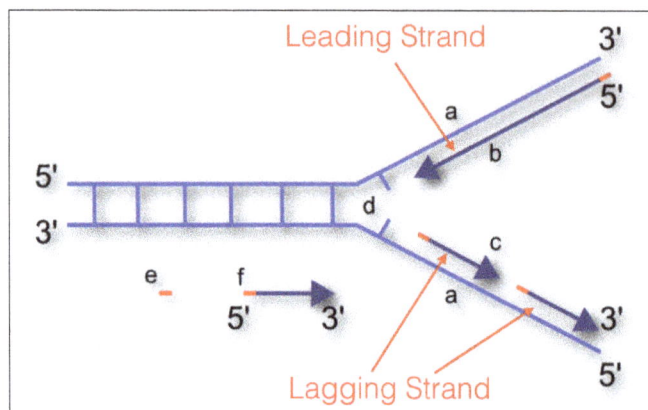

The other new strand, which runs 5′ to 3′ away from the fork, is trickier. This strand is made in fragments because, as the fork moves forward, the DNA polymerase (which is moving away from the fork) must come off and reattach on the newly exposed DNA. This tricky strand, which is made in fragments, is called the lagging strand.

The small fragments are called Okazaki fragments, named for the Japanese scientist who discovered them. The leading strand can be extended from one primer alone, whereas the lagging strand needs a new primer for each of the short Okazaki fragments.

## Maintenance and Cleanup Crew

Some other proteins and enzymes, in addition the main ones above, are needed to keep DNA replication running smoothly. One is a protein called the sliding clamp, which holds DNA polymerase III molecules in place as they synthesize DNA. The sliding clamp is a ring-shaped protein and keeps the DNA polymerase of the lagging strand from floating off when it re-starts at a new Okazaki fragment 4' end.

Topoisomerase also plays an important maintenance role during DNA replication. This enzyme prevents the DNA double helix ahead of the replication fork from getting too tightly wound as the DNA is opened up. It acts by making temporary nicks in the helix to release the tension, then sealing the nicks to avoid permanent damage.

Finally, there is a little cleanup work to do if we want DNA that doesn't contain any RNA or gaps. The RNA primers are removed and replaced by DNA through the activity of DNA polymerase I, the other polymerase involved in replication. The nicks that remain after the primers are replaced get sealed by the enzyme DNA ligase.

## DNA Replication in E. Coli

Let's see how the enzymes and proteins involved in replication work together to synthesize new DNA.

- Helicase opens up the DNA at the replication fork.

- Single-strand binding proteins coat the DNA around the replication fork to prevent rewinding of the DNA.

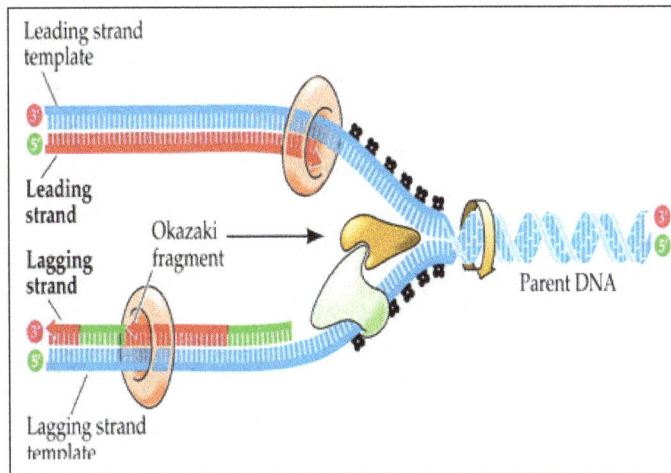

- Topoisomerase works at the region ahead of the replication fork to prevent supercoiling.

- Primase synthesizes RNA primers complementary to the DNA strand.

- DNA polymerase III extends the primers, adding on to the 3′ end, to make the bulk of the new DNA.

- RNA primers are removed and replaced with DNA by DNA polymerase I.

- The gaps between DNA fragments are sealed by DNA ligase.

## Transcription

Transcription is the first step of gene expression. During this process, the DNA sequence of a gene is copied into RNA. Before transcription can take place, the DNA double helix must unwind near the gene that is getting transcribed. The region of opened-up DNA is called a transcription bubble.

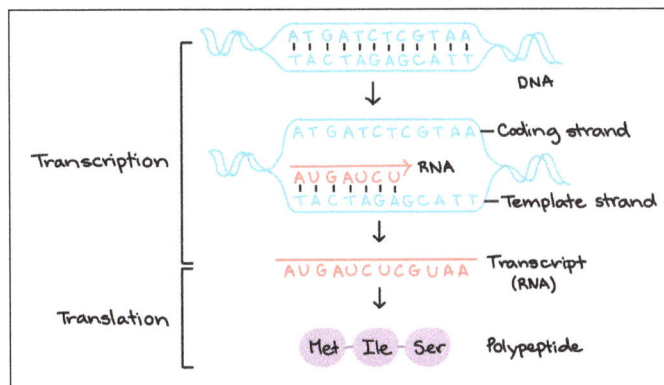

Transcription uses one of the two exposed DNA strands as a template; this strand is called the template strand. The RNA product is complementary to the template strand and is almost identical to the other DNA strand, called the nontemplate (or coding) strand. However, there is one important difference: in the newly made RNA, all of the T nucleotides are replaced with U nucleotides.

The site on the DNA from which the first RNA nucleotide is transcribed is called the, 1 site, or the initiation site. Nucleotides that come before the initiation site are given negative numbers and said to be upstream. Nucleotides that come after the initiation site are marked with positive numbers and said to be downstream. If the gene that's transcribed encodes a protein (which many genes do), the RNA molecule will be read to make a protein in a process called translation.

## RNA Polymerase

RNA polymerases are enzymes that transcribe DNA into RNA. Using a DNA template, RNA polymerase builds a new RNA molecule through base pairing. For instance, if there is a G in the DNA template, RNA polymerase will add a C to the new, growing RNA strand.

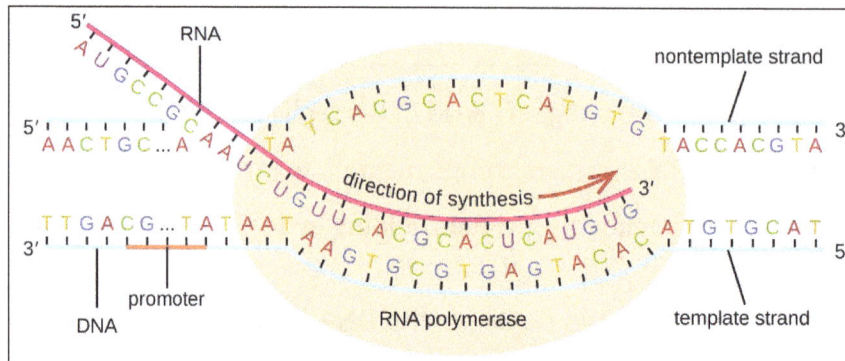

RNA polymerase always builds a new RNA strand in the 5′ to 3′ direction. That is, it can only add RNA nucleotides (A, U, C, or G) to the 3′ end of the strand. RNA polymerases are large enzymes with multiple subunits, even in simple organisms like bacteria. In addition, humans and other eukaryotes have three different kinds of RNA polymerases: I, II, and III. Each one specializes in transcribing certain classes of genes.

## Transcription Initiation

To begin transcribing a gene, RNA polymerase binds to the DNA of the gene at a region called the promoter. Basically, the promoter tells the polymerase where to "sit down" on the DNA and begin transcribing.

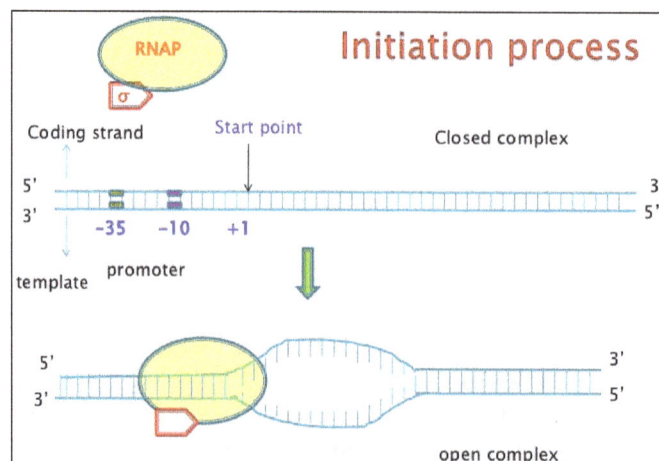

Each gene (or, in bacteria, each group of genes transcribed together) has its own promoter. A promoter contains DNA sequences that let RNA polymerase or its helper proteins attach to the DNA. Once the transcription bubble has formed, the polymerase can start transcribing.

## Promoters in Bacteria

To get a better sense of how a promoter works, let's look an example from bacteria. A typical bacterial promoter contains two important DNA sequences, the 10 and 35 elements. RNA polymerase recognizes and binds directly to these sequences. The sequences position the polymerase in the right spot to start transcribing a target gene, and they also make sure it's pointing in the right direction.

Once the RNA polymerase has bound, it can open up the DNA and get to work. DNA opening occurs at the -101010 element, where the strands are easy to separate due to the many As and Ts (which bind to each other using just two hydrogen bonds, rather than the three hydrogen bonds of Gs and Cs).

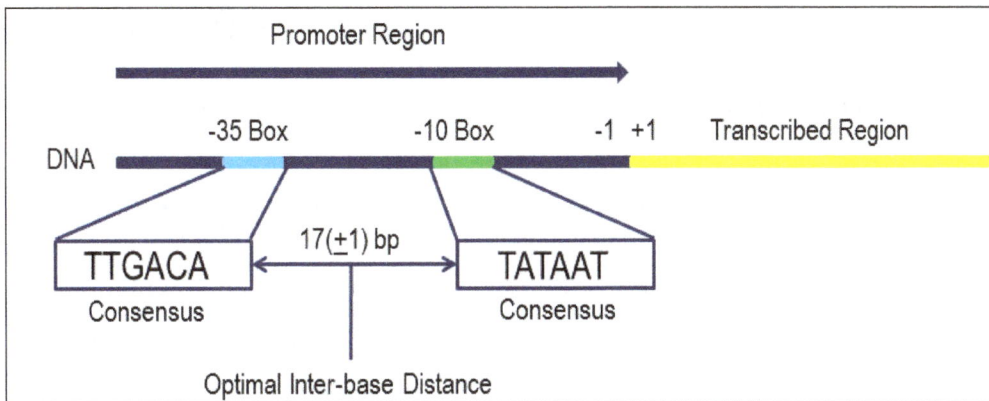

The 10 and the 35 elements get their names because they come 35 and 10 nucleotides before the initiation site. The minus signs just mean that they are before, not after, the initiation site.

## Promoters in Humans

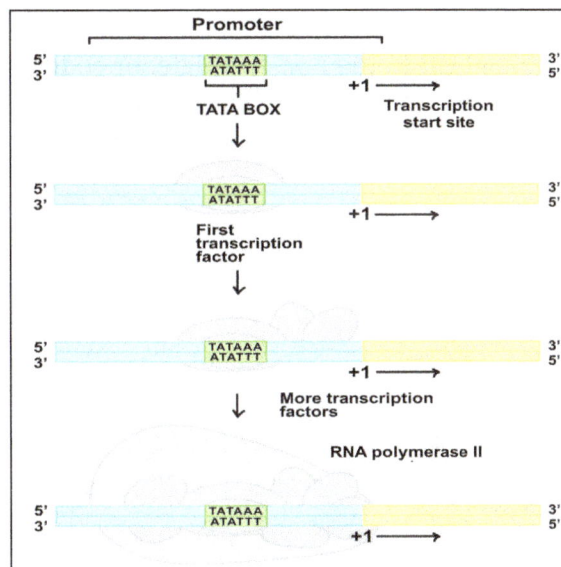

In eukaryotes like humans, the main RNA polymerase in your cells does not attach directly to promoters like bacterial RNA polymerase. Instead, helper proteins called basal (general) transcription factors bind to the promoter first, helping the RNA polymerase in your cells get a foothold on the DNA.

Many eukaryotic promoters have a sequence called a TATA box. The TATA box plays a role much like that of the -101010 element in bacteria. It's recognized by one of the general transcription factors, allowing other transcription factors and eventually RNA polymerase to bind. It also contains lots of As and Ts, which make it easy to pull the strands of DNA apart.

## Elongation

Once RNA polymerase is in position at the promoter, the next step of transcription—elongation—can begin. Basically, elongation is the stage when the RNA strand gets longer, thanks to the addition of new nucleotides.

During elongation, RNA polymerase "walks" along one strand of DNA, known as the template strand, in the 3′ to 5′ direction. For each nucleotide in the template, RNA polymerase adds a matching (complementary) RNA nucleotide to the 3′ end of the RNA strand.

Here is the reaction that adds an RNA nucleotide to the chain:

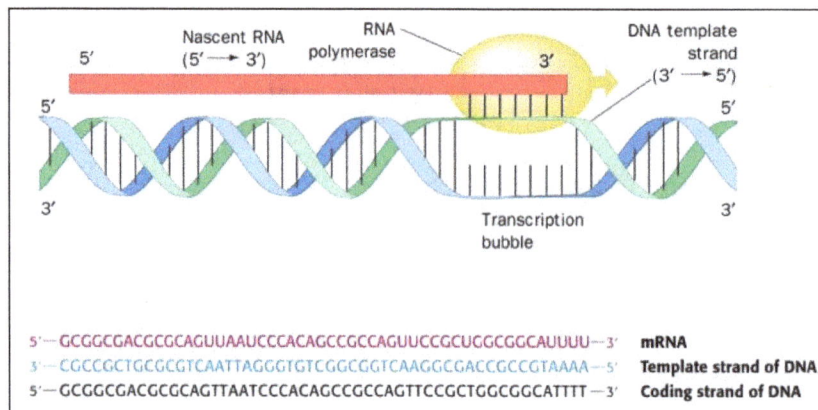

The RNA transcript is nearly identical to the non-template, or coding, strand of DNA. However, RNA strands have the base uracil (U) in place of thymine (T), as well as a slightly different sugar in the nucleotide. So, as we can see in the diagram above, each T of the coding strand is replaced with a U in the RNA transcript.

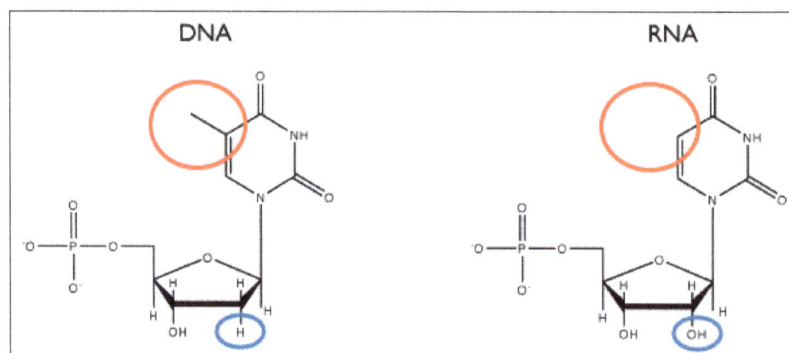

The picture below shows DNA being transcribed by many RNA polymerases at the same time, each with an RNA "tail" trailing behind it. The polymerases near the start of the gene have short RNA tails, which get longer and longer as the polymerase transcribes more of the gene.

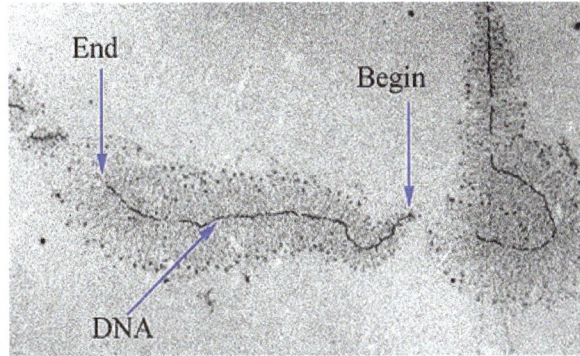

## Transcription Termination

There are two major termination strategies found in bacteria: Rho-dependent and Rho-independent.

In Rho-dependent termination, the RNA contains a binding site for a protein called Rho factor. Rho factor binds to this sequence and starts "climbing" up the transcript towards RNA polymerase.

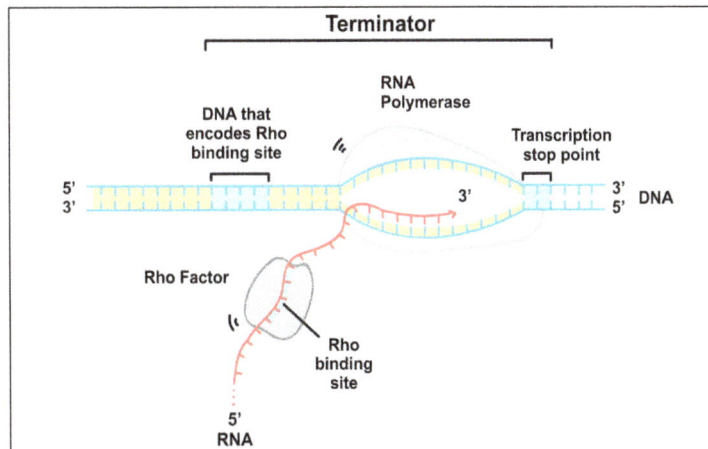

When it catches up with the polymerase at the transcription bubble, Rho pulls the RNA transcript and the template DNA strand apart, releasing the RNA molecule and ending transcription. Another sequence found later in the DNA, called the transcription stop point, causes RNA polymerase to pause and thus helps Rho catch up.

Rho-independent termination depends on specific sequences in the DNA template strand. As the RNA polymerase approaches the end of the gene being transcribed, it hits a region rich in C and G nucleotides. The RNA transcribed from this region folds back on itself, and the complementary C and G nucleotides bind together. The result is a stable hairpin that causes the polymerase to stall.

In a terminator, the hairpin is followed by a stretch of U nucleotides in the RNA, which matches up with A nucleotides in the template DNA. The complementary U-A region of the RNA transcript forms only a weak interaction with the template DNA. This, coupled with the stalled polymerase, produces enough instability for the enzyme to fall off and liberate the new RNA transcript.

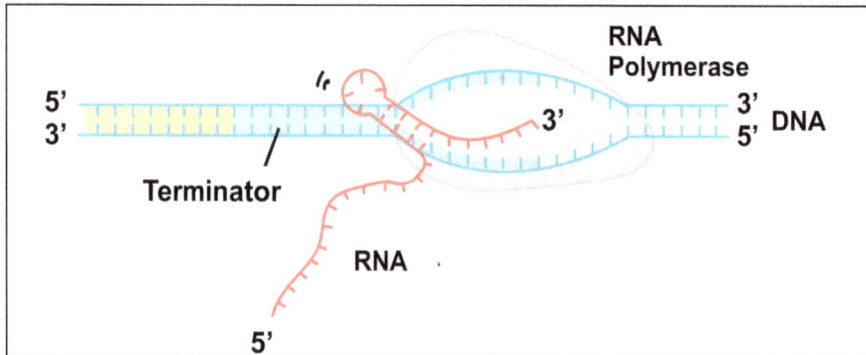

## Fate of RNA Transcript

After termination, transcription is finished. An RNA transcript that is ready to be used in translation is called a messenger RNA (mRNA). In bacteria, RNA transcripts are ready to be translated right after transcription. In fact, they're actually ready a little sooner than that: translation may start while transcription is still going on.

In the diagram, mRNAs are being transcribed from several different genes. Although transcription is still in progress, ribosomes have attached each mRNA and begun to translate it into protein. When an mRNA is being translated by multiple ribosomes, the mRNA and ribosomes together are said to form a polyribosome.

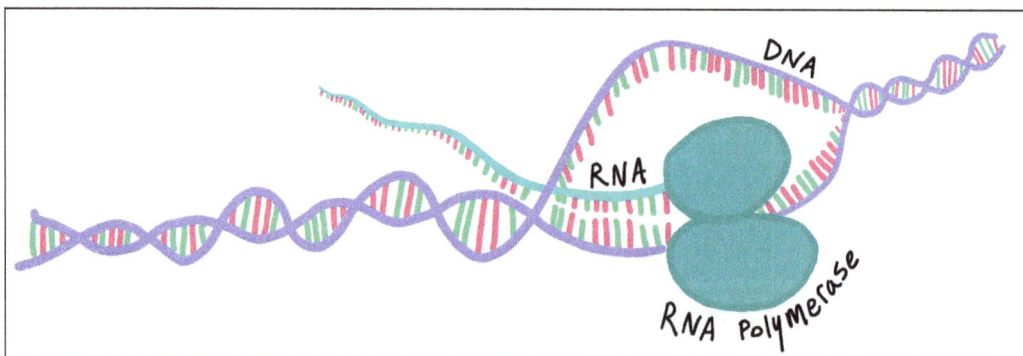

Why can transcription and translation happen simultaneously for an mRNA in bacteria? One reason is that these processes occur in the same 5′ to 3′ direction. That means one can follow or "chase" another that's still occurring. Also, in bacteria, there are no internal membrane compartments to separate transcription from translation.

The picture is different in the cells of humans and other eukaryotes. That's because transcription happens in the nucleus of human cells, while translation happens in the cytosol. Also, in eukaryotes, RNA molecules need to go through special processing steps before translation. That means translation can't start until transcription and RNA processing are fully finished.

# Proteins

Proteins are among the most abundant organic molecules in living systems and are way more diverse in structure and function than other classes of macromolecules. A single cell can contain thousands of proteins, each with a unique function. Although their structures, like their functions, vary greatly, all proteins are made up of one or more chains of amino acids.

# Protein Synthesis

### Structure of RNA

DNA alone cannot account for the expression of genes. RNA is needed to help carry out the instructions in DNA.

Like DNA, RNA is made up of nucleotide consisting of a 5-carbon sugar ribose, a phosphate group, and a nitrogenous base. However, there are three main differences between DNA and RNA:

1.   RNA uses the sugar ribose instead of deoxyribose.

2.   RNA is generally single-stranded instead of double-stranded.

3.   RNA contains uracil in place of thymine.

These differences help enzymes in the cell to distinguish DNA from RNA.

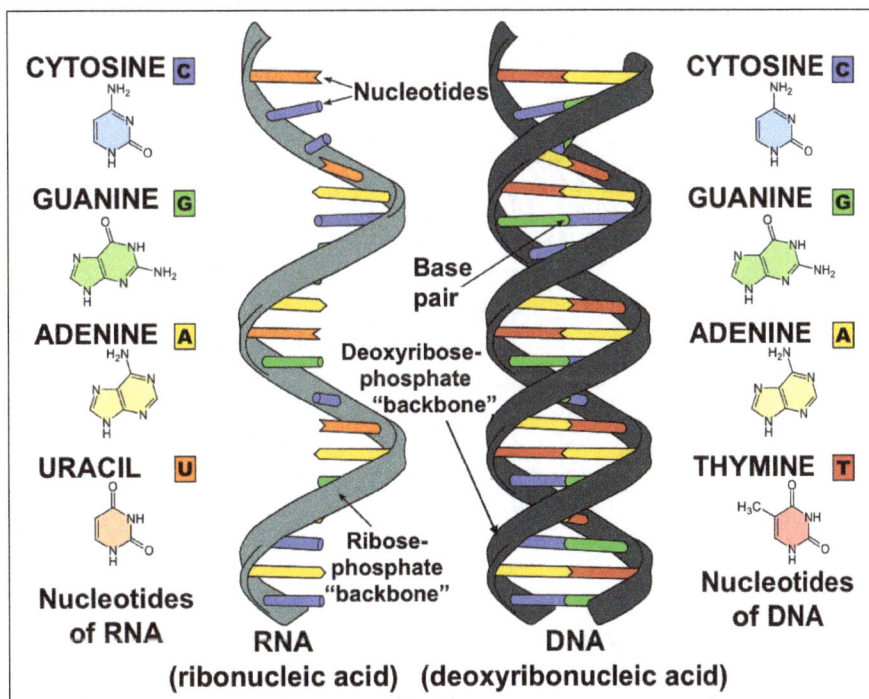

Comparison of RNA and DNA molecules.

## Types of RNA

| Type | Role |
| --- | --- |
| Messenger RNA (mRNA) | Carries information from DNA in the nucleus to ribosomes in the cytoplasm. |
| Ribosomal RNA (rRNA) | Structural component of ribosomes. |
| Transfer RNA (tRNA) | Carries amino acids to the ribosome during translation to help build an amino acid chain. |

## Central Dogma of Biology

A gene that encodes a polypeptide is expressed in two steps. In this process, information flows from DNA → RNA → protein, a directional relationship known as the central dogma of molecular biology.

## Genetic Code

The first step in decoding genetic messages is transcription, during which a nucleotide sequence is copied from DNA to RNA. The next step is to join amino acids together to form a protein. The order in which amino acids are joined together determine the shape, properties, and function of a protein.

The four bases of RNA form a language with just four nucleotide bases: adenine (A), cytosine (C), guanine (G), and uracil (U). The genetic code is read in three-base words called codons. Each codon corresponds to a single amino acid (or signals the starting and stopping points of a sequence).

Codon chart

## Transcription and Translation

In transcription, a DNA sequence is rewritten, or transcribed, into a similar RNA "alphabet." In eukaryotes, the RNA molecule must undergo processing to become a mature messenger RNA (mRNA). In translation, the sequence of the mRNA is decoded to specify the amino acid sequence of a polypeptide. The name translation reflects that the nucleotide sequence of the mRNA sequence must be translated into the completely different "language" of amino acids.

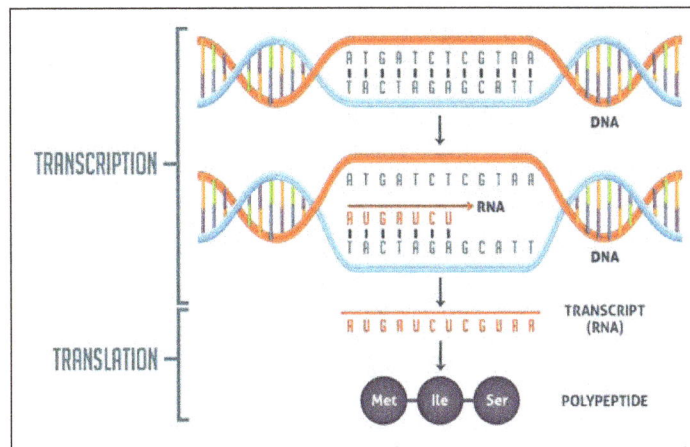

## Mutations

Sometimes cells make mistakes in copying their genetic information, causing mutations. Mutations can be irrelevant, or they effect the way proteins are made and genes are expressed.

## Substitutions

A substitution changes a single base pair by replacing one base for another.

There are three kinds of substitution mutations:

- Silent mutations do not affect the sequence of amino acids during translation.

- Nonsense mutations result in a stop codon where an amino acid should be, causing translation to stop prematurely.

- Missense mutations change the amino acid specified by a codon.

## Insertions and Deletions

An insertion occurs when one or more bases are added to a DNA sequence. A deletion occurs when one or more bases are removed from a DNA sequence. Because the genetic code is read in codons (three bases at a time), inserting or deleting bases may change the "reading frame" of the sequence. These types of mutations are called frameshift mutations.

|  | Original reading frame | New reading frame |
| --- | --- | --- |
| DNA | 3'TACTATGCCTTA-5' | 3'-TACATGCCTTA-5 |
| Mrna | 5'-AUGAUACGAAU-3' | 5'-AUGAUACGGAAU-3' |
| Codons | 5'-AUG-AUA-CGG-AAU-3' | 5'-AUG-UAC-GGA-AU-3' |
| Polypeptide | Met-lie-Arg-Asn | Met-Tyr-Gly |

As this example illustrates, a frameshift mutation changes how nucleotides are interpreted as codons beyond the point of the mutation, and this, in turn, may change the amino acid sequence.

## Common Mistakes and Misconceptions

- Amino acids are not made during protein synthesis. Some students think that the purpose of protein synthesis is to create amino acids. However, amino acids are not being made during translation, they are being used as building blocks to make proteins.

- Mutations do not always have drastic or negative effects. Often people hear the term "mutation" in the media and understand it to mean that a person will have a disease or disfigurement. Mutations are the source of genetic variety, so although some mutations are harmful, most are unnoticeable, and many are even good.

- Insertions and deletions that are multiples of three nucleotides will not cause frameshift mutations. Rather, one or more amino acids will just be added to or deleted from the protein. Insertions and deletions that are not multiples of three nucleotides, however, can dramatically alter the amino acid sequence of the protein.

DNA provides a "blueprint" for the cell structure and physiology. This refers to the fact that DNA contains the information necessary for the cell to build one very important type of molecule: the protein. Most structural components of the cell are made up, at least in part, by proteins and virtually all the functions that a cell carries out are completed with the help of proteins. One of the most important classes of proteins is enzymes, which help speed up necessary biochemical reactions that take place inside the cell. Some of these critical biochemical reactions include building larger molecules from smaller components (such as occurs during DNA replication or synthesis of microtubules) and breaking down larger molecules into smaller components (such as when harvesting chemical energy from nutrient molecules). Whatever the cellular process may be, it is almost sure to involve proteins. Just as the cell's genome describes its full complement of DNA, a cell's proteome is its full complement of proteins. Protein synthesis begins with genes. A gene is a functional segment of DNA that provides the genetic information necessary to build a protein. Each particular gene provides the code necessary to construct a particular protein. Gene expression, which transforms the information coded in a gene to a final gene product, ultimately dictates the structure and function of a cell by determining which proteins are made.

The interpretation of genes works in the following way. Recall that proteins are polymers, or chains, of many amino acid building blocks. The sequence of bases in a gene (that is, its sequence of A, T, C, G nucleotides) translates to an amino acid sequence. A triplet is a section of three DNA bases in a row that codes for a specific amino acid. Similar to the way in which the three-letter code d-o-g signals the image of a dog, the three-letter DNA base code signals the use of a particular amino acid. For example, the DNA triplet CAC (cytosine, adenine, and cytosine) specifies the amino acid valine. Therefore, a gene, which is composed of multiple triplets in a unique sequence, provides the code to build an entire protein, with multiple amino acids in the proper sequence. The mechanism by which cells turn the DNA code into a protein product is a two-step process, with an RNA molecule as the intermediate.

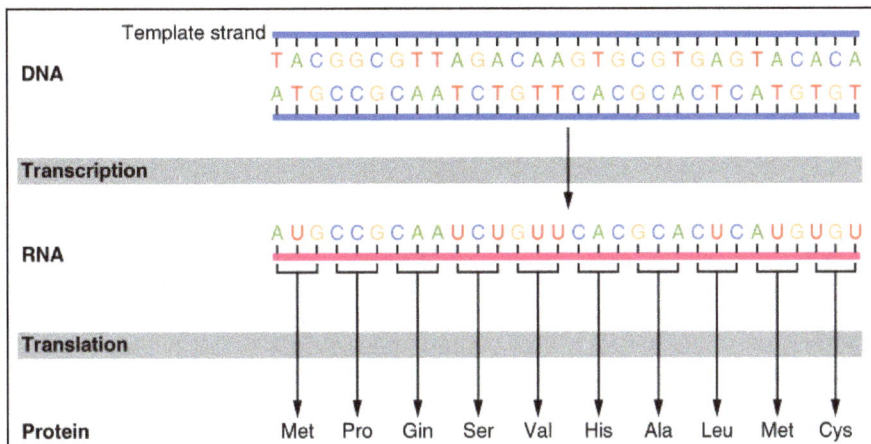

Figure: The Genetic Code.

This diagram shows the translation of RNA into proteins. A DNA template strand is shown to become an RNA strand through transcription. Then the RNA strand undergoes translation and becomes proteins.

DNA holds all of the genetic information necessary to build a cell's proteins. The nucleotide sequence of a gene is ultimately translated into an amino acid sequence of the gene's corresponding protein.

## From DNA to RNA: Transcription

DNA is housed within the nucleus, and protein synthesis takes place in the cytoplasm, thus there must be some sort of intermediate messenger that leaves the nucleus and manages protein synthesis. This intermediate messenger is messenger RNA (mRNA), a single-stranded nucleic acid that carries a copy of the genetic code for a single gene out of the nucleus and into the cytoplasm where it is used to produce proteins.

There are several different types of RNA, each having different functions in the cell. The structure of RNA is similar to DNA with a few small exceptions. For one thing, unlike DNA, most types of RNA, including mRNA, are single-stranded and contain no complementary strand. Second, the ribose sugar in RNA contains an additional oxygen atom compared with DNA. Finally, instead of the base thymine, RNA contains the base uracil. This means that adenine will always pair up with uracil during the protein synthesis process.

Gene expression begins with the process called transcription, which is the synthesis of a strand of mRNA that is complementary to the gene of interest. This process is called transcription because the mRNA is like a transcript, or copy, of the gene's DNA code. Transcription begins in a fashion somewhat like DNA replication, in that a region of DNA unwinds and the two strands separate, however, only that small portion of the DNA will be split apart. The triplets within the gene on this section of the DNA molecule are used as the template to transcribe the complementary strand of RNA. A codon is a three-base sequence of mRNA, so-called because they directly encode amino acids. Like DNA replication, there are three stages to transcription: initiation, elongation, and termination.

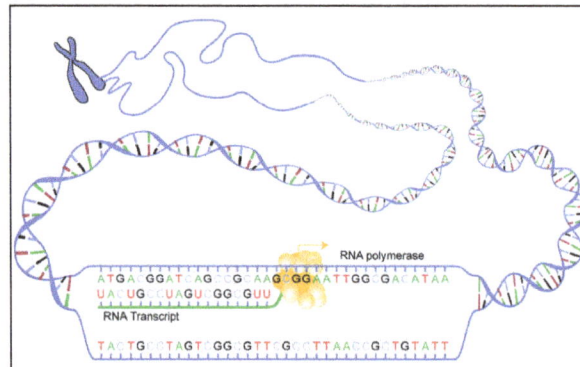

Figure: Transcription: from DNA to mRNA. In the first of the two stages of making protein from DNA, a gene on the DNA molecule is transcribed into a complementary mRNA molecule.

Stage 1: Initiation. A region at the beginning of the gene called a promoter—a particular sequence of nucleotides—triggers the start of transcription.

Stage 2: Elongation. Transcription starts when RNA polymerase unwinds the DNA segment. One strand, referred to as the coding strand, becomes the template with the genes to be coded. The polymerase then aligns the correct nucleic acid (A, C, G, or U) with its complementary base on the coding strand of DNA. RNA polymerase is an enzyme that adds new nucleotides to a growing strand of RNA. This process builds a strand of mRNA.

Stage 3: Termination. When the polymerase has reached the end of the gene, one of three specific triplets (UAA, UAG, or UGA) codes a "stop" signal, which triggers the enzymes to terminate transcription and release the mRNA transcript.

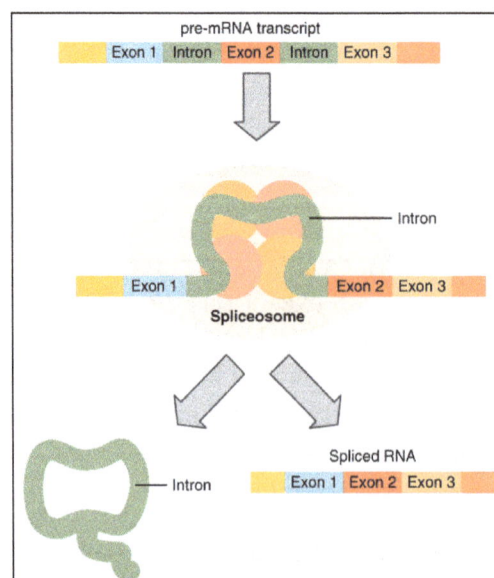

Figure: Splicing DNA. In the nucleus, a structure called a spliceosome cuts out introns (noncoding regions) within a pre-mRNA transcript and reconnects the exons.

Before the mRNA molecule leaves the nucleus and proceeds to protein synthesis, it is modified in a number of ways. For this reason, it is often called a pre-mRNA at this stage. For example, your DNA, and thus complementary mRNA, contains long regions called non-coding regions that do not code for amino acids. Their function is still a mystery, but the process called splicing removes these non-coding regions from the pre-mRNA transcript. A spliceosome—a structure composed of various proteins and other molecules—attaches to the mRNA and "splices" or cuts out the non-coding regions. The removed segment of the transcript is called an intron. The remaining exons are pasted together. An exon is a segment of RNA that remains after splicing. Interestingly, some introns that are removed from mRNA are not always non-coding. When different coding regions of mRNA are spliced out, different variations of the protein will eventually result, with differences in structure and function. This process results in a much larger variety of possible proteins and protein functions. When the mRNA transcript is ready, it travels out of the nucleus and into the cytoplasm.

## From RNA to Protein: Translation

Like translating a book from one language into another, the codons on a strand of mRNA must be translated into the amino acid alphabet of proteins. Translation is the process of synthesizing a chain of amino acids called a polypeptide. Translation requires two major aids: first, a "translator," the molecule that will conduct the translation, and second, a substrate on which the mRNA strand is translated into a new protein, like the translator's "desk." Both of these requirements are fulfilled by other types of RNA. The substrate on which translation takes place is the ribosome.

Remember that many of a cell's ribosomes are found associated with the rough ER, and carry out the synthesis of proteins destined for the Golgi apparatus. Ribosomal RNA (rRNA) is a type of RNA that, together with proteins, composes the structure of the ribosome. Ribosomes exist in the cytoplasm as two distinct components, a small and a large subunit. When an mRNA molecule is ready to be translated, the two subunits come together and attach to the mRNA. The ribosome provides a substrate for translation, bringing together and aligning the mRNA molecule with the molecular "translators" that must decipher its code.

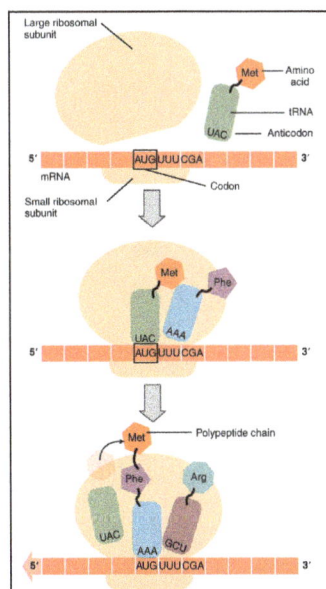

Figure: Translation from RNA to Protein.

During translation, the mRNA transcript is "read" by a functional complex consisting of the ribosome and tRNA molecules. tRNAs bring the appropriate amino acids in sequence to the growing polypeptide chain by matching their anti-codons with codons on the mRNA strand.

The other major requirement for protein synthesis is the translator molecules that physically "read" the mRNA codons. Transfer RNA (tRNA) is a type of RNA that ferries the appropriate corresponding amino acids to the ribosome, and attaches each new amino acid to the last, building the polypeptide chain one-by-one. Thus tRNA transfers specific amino acids from the cytoplasm to a growing polypeptide. The tRNA molecules must be able to recognize the codons on mRNA and match them with the correct amino acid. The tRNA is modified for this function. On one end of its structure is a binding site for a specific amino acid. On the other end is a base sequence that matches the codon specifying its particular amino acid? This sequence of three bases on the tRNA molecule is called an anticodon. For example, a tRNA responsible for shuttling the amino acid glycine contains a binding site for glycine on one end. On the other end it contains an anticodon that complements the glycine codon (GGA is a codon for glycine, and so the tRNAs anticodon would read CCU). Equipped with its particular cargo and matching anticodon, a tRNA molecule can read its recognized mRNA codon and bring the corresponding amino acid to the growing chain.

Much like the processes of DNA replication and transcription, translation consists of three main stages: initiation, elongation, and termination. Initiation takes place with the binding of a ribosome to an mRNA transcript. The elongation stage involves the recognition of a tRNA anticodon with the next mRNA codon in the sequence. Once the anticodon and codon sequences are bound, the tRNA presents its amino acid cargo and the growing polypeptide strand is attached to this next amino acid. This attachment takes place with the assistance of various enzymes and requires energy. The tRNA molecule then releases the mRNA strand, the mRNA strand shifts one codon over in the ribosome, and the next appropriate tRNA arrives with its matching anticodon. This process continues until the final codon on the mRNA is reached which provides a "stop" message that signals termination of translation and triggers the release of the complete, newly synthesized protein. Thus, a gene within the DNA molecule is transcribed into mRNA, which is then translated into a protein product.

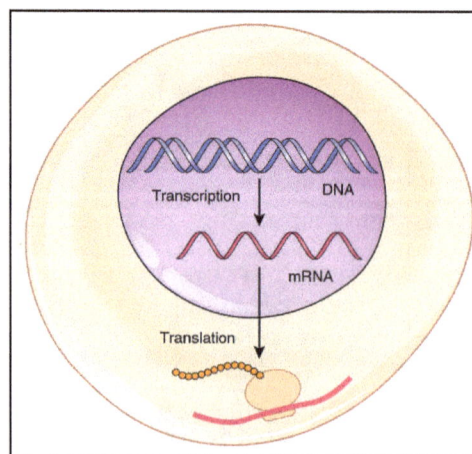

Figure: From DNA to Protein: Transcription through Translation.

Transcription within the cell nucleus produces an mRNA molecule, which is modified and then sent into the cytoplasm for translation. The transcript is decoded into a protein with the help of a ribosome and tRNA molecules.

Commonly, an mRNA transcription will be translated simultaneously by several adjacent ribosomes. This increases the efficiency of protein synthesis. A single ribosome might translate an mRNA molecule in approximately one minute; so multiple ribosomes aboard a single transcript could produce multiple times the number of the same protein in the same minute. A polyribosome is a string of ribosomes translating a single mRNA strand.

## References

- Meiosis, cell-biology, science: britannica.com, Retrieved 3 March, 2019

- Molecular-mechanism-of-dna-replication, dna-replication, dna-as-the-genetic-material, biology, science: khanacademy.org, Retrieved 13 January, 2019

- Stages-of-transcription, transcription-of-dna-into-rna, gene-expression-central-dogma, biology, science: khanacademy.org, Retrieved 18 May, 2019

- Introduction-to-proteins-and-amino-acids, proteins-and-amino-acids, macromolecules, biology, science: khanacademy.org, Retrieved 8 February, 2019

- Hs-rna-and-protein-synthesis-review, hs-rna-and-protein-synthesis, hs-molecular-genetics, biology, science: khanacademy.org, Retrieved 27 June, 2019

- 3-4-protein-synthesis, anatomyandphysiology: opentextbc.ca, Retrieved 14 April, 2019

# Chapter 3

# Techniques in Molecular Biology

Molecular biology makes use of diverse techniques. Some of them are molecular cloning, polymerase chain reaction, gel electrophoresis, blotting and DNA microarray. The diverse applications of these techniques in the field of molecular biology have been thoroughly discussed in this chapter.

## Molecular Cloning

Molecular cloning refers to the isolation of a DNA sequence from any species (often a gene), and its insertion into a vector for propagation, without alteration of the original DNA sequence. Once isolated, molecular clones can be used to generate many copies of the DNA for analysis of the gene sequence, and to express the resulting protein for the study or utilization of the protein's function. The clones can also be manipulated and mutated in vitro to alter the expression and function of the protein.

The basic cloning workflow includes four steps:

1. Isolation of target DNA fragments (often referred to as inserts).

2. Ligation of inserts into an appropriate cloning vector, creating recombinant molecules (e.g., plasmids).

3. Transformation of recombinant plasmids into bacteria or other suitable host for propagation.

4. Screening/selection of hosts containing the intended recombinant plasmid.

These four ground-breaking steps were carefully pieced together and performed by multiple laboratories, beginning in the late 1960s and early 1970s.

### Foundation of Molecular Cloning

### Cutting

Recombinant DNA technology first emerged in the late 1960s, with the discovery of enzymes that could specifically cut and join double-stranded DNA molecules. In fact, as early as 1952, two groups independently observed that bacteria encoded a "restriction factor" that prevented bacteriophages from growing within certain hosts. But the nature of the factor was not discovered until 1968, when Arber and Linn succeeded in isolating an enzyme, termed a restriction factor that selectively cut exogenous DNA, but not bacterial DNA. These studies also identified a methylase enzyme that protected the bacterial DNA from restriction enzymes.

Shortly after Arber and Linn's discovery, Smith extended and confirmed these studies by isolating a restriction enzyme from *Haemophilus influenza*. He demonstrated that the enzyme selectively cut DNA in the middle of a specific 6 base-pair stretch of DNA; one characteristic of certain restriction enzymes is their propensity to cut the DNA substrate in or near specific, often palindromic, "recognition" sequences.

The full power of restriction enzymes was not realized until restriction enzymes and gel electrophoresis were used to map the Simian Virus 40 (SV40) genome. For these seminal findings, Werner Arber, Hamilton Smith, and Daniel Nathans shared the 1978 Nobel Prize in Medicine.

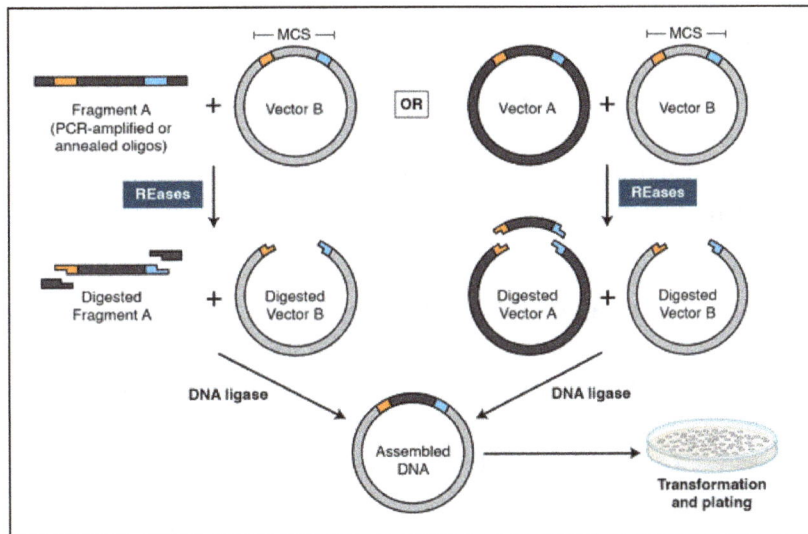

Figure: Traditional Cloning Workflow.

Using PCR, restriction sites are added to both ends of a dsDNA, which is then digested by the corresponding restriction enzymes. The cleaved DNA can then be ligated to a plasmid vector possessing compatible ends. DNA fragments can also be moved from one vector into another by digesting with Rises and ligating with compatible ends of the target vector. Assembled construct can then be transformed into Escherichia coli (E. coli).

## Assembling

Much like the discovery of enzymes that cut DNA, the discovery of an enzyme that could join DNA was preceded by earlier, salient observations. In the early 1960s, two groups discovered that genetic recombination could occur though the breakage and ligation of DNA molecules, closely followed by the observation that linear bacteriophage DNA is rapidly converted to covalently closed circles after infection of the host. Just two years later, five groups independently isolated DNA ligases and demonstrated their ability to assemble two pieces of DNA.

Not long after the discovery of restriction enzymes and DNA ligases, the first recombinant DNA molecule was made. In 1972, Berg separately cut and ligated a piece of lambda bacteriophage DNA or the *E. coli* galactose operon with SV40 DNA to create the first recombinant DNA molecules. These studies pioneered the concept that, because of the universal nature of DNA, DNA from any species could be joined together. In 1980, Paul Berg shared the Nobel Prize in Chemistry with Walter Gilbert and Frederick Sanger (the developers of DNA sequencing), for "his fundamental studies of the biochemistry of nucleic acids, with particular regard to recombinant DNA."

## Transformation

Recombinant DNA technology would be severely limited, and molecular cloning impossible, without the means to propagate and isolate the newly constructed DNA molecule. The ability to transform bacteria, or induce the uptake, incorporation and expression of foreign genetic material, was first demonstrated by Griffith when he transformed a non-lethal strain of bacteria into a lethal strain by mixing the non-lethal strain with heat-inactivated lethal bacteria. However, the nature of the "transforming principle" that conveyed lethality was not understood until 1944. In the same year, Avery, Macleod and McCarty demonstrated that DNA, and not protein, was responsible for inducing the lethal phenotype.

Initially, it was believed that the common bacterial laboratory strain, *E. coli*, was refractory to transformation, until Mandel and Higa demonstrated that treatment of *E. coli* with calcium chloride induced the uptake of bacteriophage DNA. Cohen applied this principle, in 1972, when he pioneered the transformation of bacteria with plasmids to confer antibiotic resistance on the bacteria.

The ultimate experiment digestion, ligation and transformation of a recombinant DNA molecule was executed by Boyer, Cohen and Chang in 1973, when they digested the plasmid pSC101 with EcoRI, ligated the linearized fragment to another enzyme-restricted plasmid and transformed the resulting recombinant molecule into *E. coli*, conferring tetracycline resistance on the bacteria, thus laying the foundation for most recombinant DNA work since.

## Building on the Groundwork

While scientists had discovered and applied all of the basic principles for creating and propagating recombinant DNA in bacteria, the process was inefficient. Restriction enzyme preparations were unreliable due to non-standardized purification procedures, plasmids for cloning were cumbersome, difficult to work with and limited in number, and experiments were limited by the amount of insert DNA that could be isolated. Research over the next few decades led to improvements in the techniques and tools available for molecular cloning.

### Early Vector Design

1. Development of the first standardized vector- Scientists working in Boyer's lab recognized the need for a general cloning plasmid, a compact plasmid with unique restriction sites for cloning in foreign DNA and the expression of antibiotic resistance genes for selection of transformed bacteria. In 1977, they described the first vector designed for cloning purposes, pBR322. This vector was small, 4 kilobases in size, and had two antibiotic resistance genes for selection.

2. Vectors with on-board screening and higher yields, Although antibiotic selection prevented non-transformed bacteria from growing, plasmids that re-ligated without insert DNA fragments (self-ligation) could still confer antibiotic resistance on bacteria. Therefore, finding the correct bacterial clones containing the desired recombinant DNA molecule could be time-consuming.

Vieira and Messing devised a screening tool to identify bacterial colonies containing plasmids with DNA inserts. Based upon the pBR322 plasmid, they created the series of pUC plasmids, which contained a "blue/white screening" system. Placement of a multiple cloning site (MCS) containing several unique restriction sites within the LacZ´ gene allowed researchers to screen for bacterial colonies containing plasmids with the foreign DNA insert. When bacteria were plated on the correct media, white colonies contained plasmids with inserts, while blue colonies contained plasmids with no inserts. pUC plasmids had an additional advantage over existing vectors; they contained a mutation that resulted in higher copy numbers, therefore increasing plasmid yields.

## Improving Restriction Digests

Early work with restriction enzymes was hampered by the purity of the enzyme preparation and a lack of understanding of the buffer requirements for each enzyme. In 1975, New England Bio labs (NEB) became the first company to commercialize restriction enzymes produced from a recombinant source. This enabled higher yields, improved purity, lot-to-lot consistency and lower pricing. Currently, over 4,000 restriction enzymes, recognizing over 300 different sequences, have been discovered by scientists across the globe. NEB currently supplies over 230 of these specificities.

NEB was also one of the first companies to develop a standardized four-buffer system, and to characterize all of its enzyme activities in this buffer system. This led to a better understanding of how to conduct a double digest, or the digestion of DNA with two enzymes simultaneously. Later research led to the development of one-buffer systems, which are compatible with the most common restriction enzymes.

With the advent of commercially available restriction enzyme libraries with known sequence specificities, restriction enzymes became a powerful tool for screening potential recombinant DNA clones. The "diagnostic digest" was, and still is, one of the most common techniques used in molecular cloning.

## Vector and Insert Preparation

Cloning efficiency and versatility were also improved by the development of different techniques for preparing vectors prior to ligation. Alkaline phosphatases were isolated that could remove the 3´ and 5´ phosphate groups from the ends of DNA and RNA. It was soon discovered that treatment of vectors with Calf-Intestinal Phosphatase (CIP) dephosphorylated DNA ends and prevented self-ligation of the vector, increasing recovery of plasmids with insert.

The CIP enzyme proved difficult to inactivate, and any residual activity led to dephosphorylation of insert DNA and inhibition of the ligation reaction. The discovery of the heat-labile alkaline phosphatases, such as recombinant Shrimp Alkaline Phosphatase (rSAP) and Antarctic Phosphatase (AP), decreased the steps and time involved, as a simple shift in temperature inactivates the enzyme prior to the ligation step.

DNA sequencing arrives: DNA sequencing was developed in the late 1970s when two competing methods were devised. Maxam and Gilbert developed the "chemical sequencing method," which relied on chemical modification of DNA and subsequent cleavage at specific bases. At the same time, Sanger and colleagues published on the "chain-termination method", which became the method used by most researchers. The Sanger method quickly became automated, and the first automatic sequencers were sold in 1987.

The ability to determine the sequence of a stretch of DNA enhanced the reliability and versatility of molecular cloning. Once cloned, scientists could sequence clones to definitively identify the correct recombinant molecule, identify new genes or mutations in genes, and easily design oligonucleotides based on the known sequence for additional experiments.

The impact of the polymerase chain reaction: One of the problems in molecular cloning in the early years was obtaining enough insert DNA to clone into the vector. In 1983, Mullis devised a technique that solved this problem and revolutionized molecular cloning. He amplified a stretch of target DNA by using opposing primers to amplify both complementary strands of DNA, simultaneously. Through cycles of denaturation, annealing and polymerization, he showed he could exponentially amplify a single copy of DNA. The polymerase chain reaction, or PCR, made it possible to amplify and clone genes from previously inadequate quantities of DNA. For this discovery, Kary Mullis shared the 1993 Nobel Prize in Chemistry "for contributions to the developments of methods within DNA-based chemistry".

In 1970, Temin and Baltimore independently discovered reverse transcriptase in viruses, an enzyme that converts RNA into DNA. Shortly after PCR was developed, reverse transcription was coupled with PCR (RT-PCR) to allow cloning of messenger RNA (mRNA). Reverse transcription was used to create a DNA copy (cDNA) of mRNA that was subsequently amplified by PCR to create an insert for ligation. For their discovery of the enzyme, Howard Temin and David Baltimore were awarded the 1975 Nobel Prize in Medicine and Physiology, which they shared with Renato Dulbecco.

Cloning of PCR products: The advent of PCR meant that researchers could now clone genes and DNA segments with limited knowledge of amplicon sequence. However, there was little consensus as to the optimal method of PCR product preparation for efficient ligation into cloning vectors.

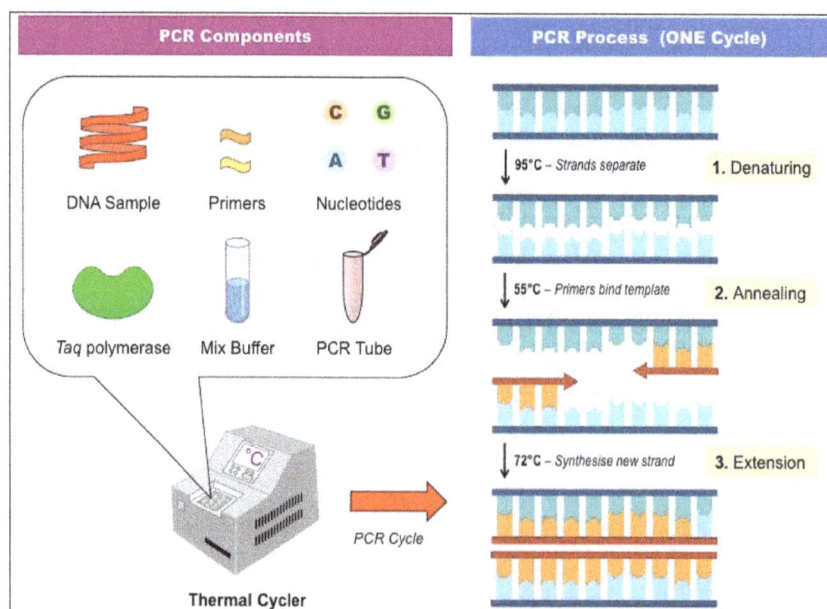

Figure: Overview of PCR.

Several different methods were initially used for cloning PCR products. The simplest, and still the most common, method for cloning PCR products is through the introduction of restriction sites onto the ends of the PCR product. This allows for direct, directional cloning of the insert into the vector after restriction digestion. Blunt-ended cloning was developed to directly ligate PCR products generated by polymerases that produced blunt ends, or inserts engineered to have restriction sites that left blunt ends once the insert was digested. This was useful in cloning DNA fragments that did not contain restriction sites compatible with the vector.

Shortly after the introduction of PCR, overlap extension PCR was introduced as a method to assemble PCR products into one contiguous DNA sequence. In this method, the DNA insert is amplified by PCR using primers that generate a PCR product containing overlapping regions with the vector. The vector and insert are then mixed, denatured and annealed, allowing hybridization of the insert to the vector. A second round of PCR generates recombinant DNA molecules of insert-containing vector. Overlap extension PCR enabled researchers to piece together large genes that could not easily be amplified by traditional PCR methods. Overlap extension PCR was also used to introduce mutations into gene sequences.

## Development of Specialized Cloning Techniques

In an effort to further improve the efficiency of molecular cloning, several specialized tools and techniques were developed that exploited the properties of unique enzymes.

1. TA Cloning- One approach took advantage of a property of Taq DNA Polymerase, the first heat-stable polymerase used for PCR. During amplification, Taq adds a single 3′ dA nucleotide to the end of each PCR product. The PCR product can be easily ligated into a vector that has been cut and engineered to contain single T residues on each strand. Several companies have marketed the technique and sell kits containing cloning vectors that are already linearized and "tailed".

2. LIC- Ligation independent cloning (LIC), as its name implies, allows for the joining of DNA molecules in the absence of DNA ligase. LIC is commonly performed with T4 DNA Polymerase, which is used to generate single-stranded DNA overhangs, >12 nucleotides long, onto both the linearized vector DNA and the insert to be cloned. When mixed together, the vector and insert anneal through the long stretch of compatible ends. The length of the compatible ends is sufficient to hold the molecule together in the absence of ligase, even during transformation. Once transformed, the gaps are repaired in vivo. There are several different commercially available products for LIC.

USER cloning was first developed in the early 1990s as a restriction enzyme- and ligase-independent cloning method. When first conceived, the method relied on using PCR primers that contained a ~12 nucleotide 5′ tail, in which at least four deoxythymidine bases had been substituted with deoxyuridines. The PCR product was treated with uracil DNA glycosidase (UDG) and Endonuclease VIII, which excises the uracil bases and leaves a 3′ overlap that can be annealed to a similarly treated vector. NEB sells the USER enzyme for ligase and restriction enzyme independent cloning reactions.

## Future Trends

Molecular cloning has progressed from the cloning of a single DNA fragment to the assembly of multiple DNA components into a single contiguous stretch of DNA. New and emerging technologies seek to transform cloning into a process that is as simple as arranging "blocks" of DNA next to each other.

## DNA Assembly Methods

Many new, elegant technologies allow for the assembly of multiple DNA fragments in a one-tube reaction. The advantages of these technologies are that they are standardized, seamless and mostly sequence independent. In addition, the ability to assemble multiple DNA fragments in one tube turns a series of previously independent restriction/ligation reactions into a streamlined, efficient procedure.

Different techniques and products for gene assembly include SLIC (Sequence and Ligase Independent Cloning), Gibson Assembly (NEB), GeneArt Seamless Cloning (Life Technologies) and Gateway Cloning (Invitrogen).

In DNA assembly, blocks of DNA to be assembled are PCR amplified. Then, the DNA fragments to be assembled adjacent to one another are engineered to contain blocks of complementary sequences that will be ligated together. These could be compatible cohesive ends, such as those used for Gibson Assembly, or regions containing recognition sites for site-specific recombines (Gateway). The enzyme used for DNA ligation will recognize and assemble each set of compatible regions, creating a single, contiguous DNA molecule in one reaction.

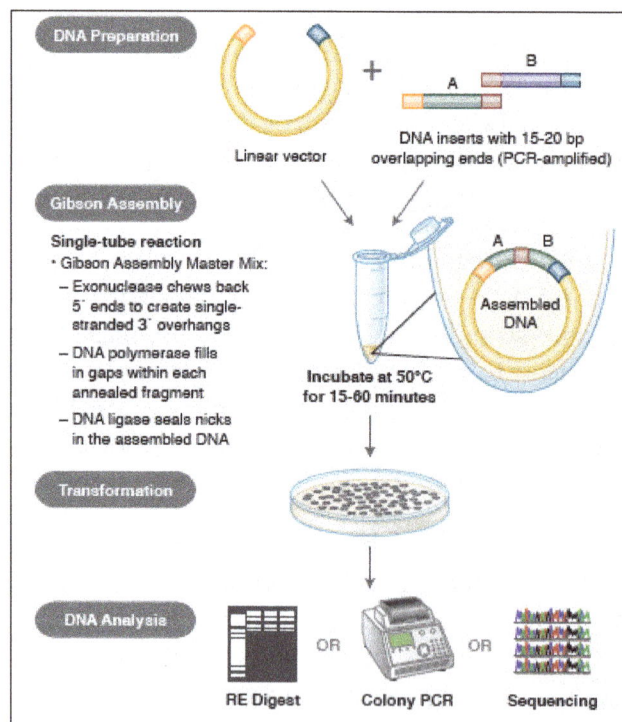

Figure: Overview of the Gibson Assembly Cloning Method.

## Synthetic Biology

DNA synthesis is an area of synthetic biology that is currently revolutionizing recombinant DNA technology. Although a complete gene was first synthesized in vitro in 1972, DNA synthesis of large DNA molecules did not become a reality until the early 2000s, when researchers began synthesizing whole genomes in vitro. These early experiments took years to complete, but technology is accelerating the ability to synthesize large DNA molecules.

# Polymerase Chain Reaction

Polymerase chain reaction (PCR) is a common laboratory technique used to make many copies (millions or billions) of a particular region of DNA. This DNA region can be anything the experimenter is interested in. For example, it might be a gene whose function a researcher wants to understand, or a genetic marker used by forensic scientists to match crime scene DNA with suspects.

Typically, the goal of PCR is to make enough of the target DNA region that it can be analyzed or used in some other way. For instance, DNA amplified by PCR may be sent for sequencing, visualized by gel electrophoresis, or cloned into a plasmid for further experiments.

PCR is used in many areas of biology and medicine, including molecular biology research, medical diagnostics, and even some branches of ecology.

## Taq Polymerase

Like DNA replication in an organism, PCR requires a DNA polymerase enzyme that makes new strands of DNA, using existing strands as templates. The DNA polymerase typically used in PCR is called Taq polymerase, after the heat-tolerant bacterium from which it was isolated (Thermus aquaticus).

*T. aquaticus* lives in hot springs and hydrothermal vents. Its DNA polymerase is very heat-stable and is most active around 70°C (a temperature at which a human or *E. coli* DNA polymerase would be nonfunctional). This heat-stability makes Taq polymerase ideal for PCR. High temperature is used repeatedly in PCR to denature the template DNA, or separate its strands.

## PCR Primers

Like other DNA polymerases, *Taq* polymerase can only make DNA if it's given a primer, a short sequence of nucleotides that provides a starting point for DNA synthesis. In a PCR reaction, the experimenter determines the region of DNA that will be copied, or amplified, by the primers she or he chooses.

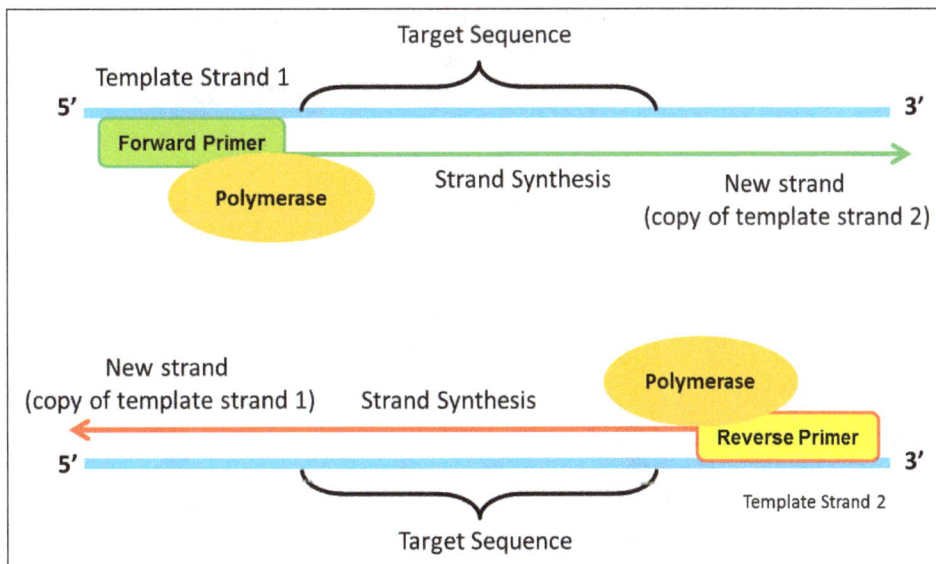

PCR primers are short pieces of single-stranded DNA, usually around 20 nucleotides in length. Two primers are used in each PCR reaction, and they are designed so that they flank the target region (region that should be copied). That is, they are given sequences that will make them bind to opposite strands of the template DNA, just at the edges of the region to be copied. The primers bind to the template by complementary base pairing.

When the primers are bound to the template, they can be extended by the polymerase, and the region that lies between them will get copied.

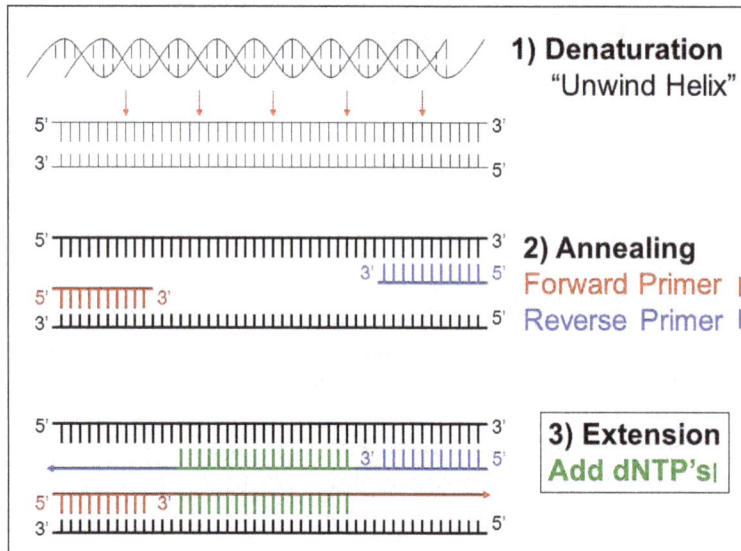

Both primers, when bound, point "inward" – that is, in the 5′ to 3′ direction towards the region to be copied. Like other DNA polymerases, Taq polymerase can only synthesize DNA in the 5′ to 3′ direction. When the primers are extended, the region that lies between them will thus be copied.

## Steps of PCR

The key ingredients of a PCR reaction are Taq polymerase, primers, template DNA, and nucleotides (DNA building blocks). The ingredients are assembled in a tube, along with cofactors needed by the enzyme, and are put through repeated cycles of heating and cooling that allow DNA to be synthesized.

The basic steps are:

1.  Denaturation (96°C): Heat the reaction strongly to separate, or denature, the DNA strands. This provides single-stranded template for the next step.

2.  Annealing (55 - 65°C): Cool the reaction so the primers can bind to their complementary sequences on the single-stranded template DNA.

3.  Extension (72°C): Raise the reaction temperatures so Taq polymerase extends the primers, synthesizing new strands of DNA.

This cycle repeats 25 - 35 times in a typical PCR reaction, which generally takes 2 - 4 hours, depending on the length of the DNA region being copied. If the reaction is efficient (works well), the target region can go from just one or a few copies to billions.

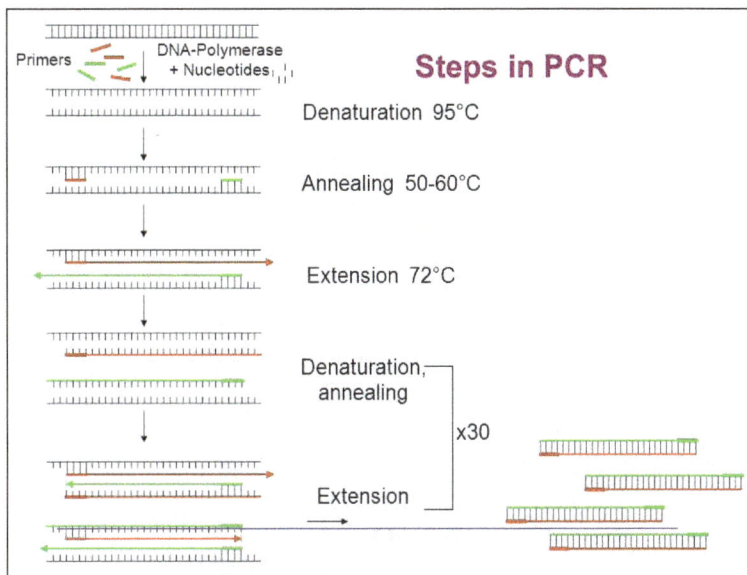

That's because it's not just the original DNA that's used as a template each time. Instead, the new DNA that's made in one round can serve as a template in the next round of DNA synthesis. There are many copies of the primers and many molecules of Taq polymerase floating around in the reaction, so the number of DNA molecules can roughly double in each round of cycling. This pattern of exponential growth is shown in the image below:

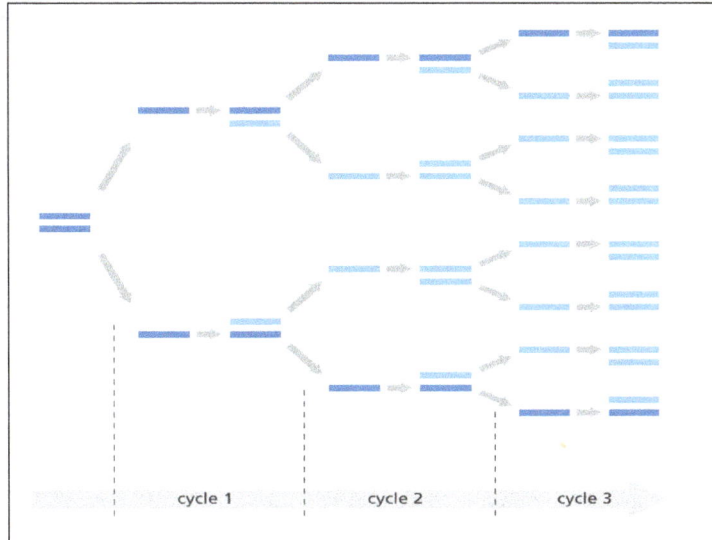

## Using Gel Electrophoresis to Visualize the Results of PCR

The results of a PCR reaction are usually visualized (made visible) using gel electrophoresis. Gel electrophoresis is a technique in which fragments of DNA are pulled through a gel matrix by an electric current, and it separates DNA fragments according to size. A standard, or DNA ladder, is typically included so that the size of the fragments in the PCR sample can be determined.

DNA fragments of the same length form a "band" on the gel, which can be seen by eye if the gel is stained with a DNA-binding dye. For example, a PCR reaction producing a 400 base pair (bp) fragment would look like this on a gel:

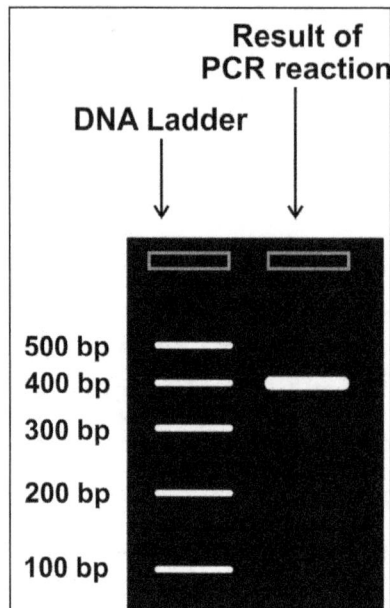

A DNA band contains many, many copies of the target DNA region, not just one or a few copies. Because DNA is microscopic, lots of copies of it must be present before we can see it by eye. This is a big part of why PCR is an important tool: it produces enough copies of a DNA sequence that we can see or manipulate that region of DNA.

## Applications of PCR

Using PCR, a DNA sequence can be amplified millions or billions of times, producing enough DNA copies to be analyzed using other techniques. For instance, the DNA may be visualized by gel electrophoresis, sent for sequencing, or digested with restriction enzymes and cloned into a plasmid.

PCR is used in many research labs, and it also has practical applications in forensics, genetic testing, and diagnostics. For instance, PCR is used to amplify genes associated with genetic disorders from the DNA of patients (or from fetal DNA, in the case of prenatal testing). PCR can also be used to test for a bacterium or DNA virus in a patient's body: if the pathogen is present, it may be possible to amplify regions of its DNA from a blood or tissue sample.

## Sample Problem: PCR in Forensics

Suppose that you are working in a forensics lab. You have just received a DNA sample from a hair left at a crime scene, along with DNA samples from three possible suspects. Your job is to examine a particular genetic marker and see whether any of the three suspects matches the hair DNA for this marker.

The marker comes in two alleles, or versions. One contains a single repeat (brown region below), while the other contains two copies of the repeat. In a PCR reaction with primers that flank the repeat region, the first allele produces a 200 bp DNA fragment, while the second produces a 300 bp DNA fragment:

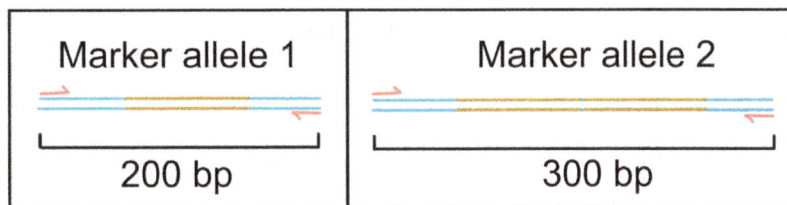

| Marker allele 1 | Marker allele 2 |
|---|---|
| 200 bp | 300 bp |

You perform PCR on the four DNA samples and visualize the results by gel electrophoresis, as shown below:

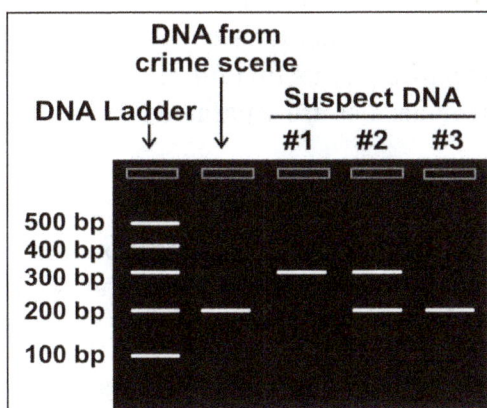

## PCR and Forensics

In real forensic tests of DNA from a crime scene, technicians would do an analysis conceptually similar to the one in the example above. However, a number of different markers (not just the single marker in the example) would be compared between the crime scene DNA and the suspects' DNA.

Also, the markers used in a typical forensic analysis don't come in just two different forms. Instead, they're highly polymorphic (poly = many, morph = form). That is, they come in many alleles that vary in tiny increments of length.

The most commonly used type of markers in forensics, called short tandem repeats (STRs), consist of many repeating copies of the same short nucleotide sequence (typically, 2 to 5 nucleotides long). One allele of an STR might have 20 repeats, while another might have 18, and another just 10, 1' end.

By examining multiple markers, each of which comes in many allele forms, forensic scientists can build a unique genetic "fingerprint" from a DNA sample. In a typical STR analysis using 13 markers, the odds of a false positive (two people having the same DNA "fingerprint") are less than 1 in 10 billion.

Although we may think of DNA evidence being used to convict criminals, it has played a crucial role in exonerating falsely accused people (including some who had been jailed for many years). Forensic analysis is also used to establish paternity and to identify human remains from disaster scenes.

# Gel Electrophoresis

Gel electrophoresis is a technique commonly used in laboratories to separate charged molecules like DNA, RNA and proteins according to their size. Charged molecules move through a gel when an electric current is passed across it.

An electric current is applied across the gel so that one end of the gel has a positive charge and the other end has a negative charge. The movement of charged molecules is called migration. Molecules migrate towards the opposite charge. A molecule with a negative charge will therefore be pulled towards the positive end (opposites attract).

The gel consists of a permeable matrix, a bit like a sieve, through which molecules can travel when an electric current is passed across it. Smaller molecules migrate through the gel more quickly and therefore travel further than larger fragments that migrate more slowly and therefore will travel a shorter distance. As a result the molecules are separated by size.

## Gel Electrophoresis and DNA

Electrophoresis enables you to distinguish DNA fragments of different lengths. DNA is negatively charged, therefore, when an electric current is applied to the gel, DNA will migrate towards the positively charged electrode.

Shorter strands of DNA move more quickly through the gel than longer strands resulting in the fragments being arranged in order of size. The use of dyes, fluorescent tags or radioactive labels enables the DNA on the gel to be seen after they have been separated. They will appear as bands on the gel.

A DNA marker with fragments of known lengths is usually run through the gel at the same time as the samples. By comparing the bands of the DNA samples with those from the DNA marker, you can work out the approximate length of the DNA fragments in the samples.

## How is Gel Electrophoresis Carried Out

## Preparing the Gel

Agarose gels are typically used to visualise fragments of DNA. The concentration of agarose used to make the gel depends on the size of the DNA fragments you are working with. The higher the agarose concentration, the denser the matrix and vice versa. Smaller fragments of DNA are separated on higher concentrations of agarose whilst larger molecules require a lower concentration of agarose.

To make a gel, agarose powder is mixed with an electrophoresis buffer and heated to a high temperature until all of the agarose powder has melted. The molten gel is then poured into a gel casting tray and a "comb" is placed at one end to make wells for the sample to be pipetted into.

Once the gel has cooled and solidified (it will now be opaque rather than clear) the comb is removed. Many people now use pre-made gels. The gel is then placed into an electrophoresis tank and electrophoresis buffer is poured into the tank until the surface of the gel is covered. The buffer conducts the electric current. The type of buffer used depends on the approximate size of the DNA fragments in the sample.

## Preparing the DNA for Electrophoresis

A dye is added to the sample of DNA prior to electrophoresis to increase the viscosity of the sample which will prevent it from floating out of the wells and so that the migration of the sample through the gel can be seen. A DNA marker (also known as a size standard or a DNA ladder) is loaded into the first well of the gel. The fragments in the marker are of a known length so can be used to help approximate the size of the fragments in the samples.

The prepared DNA samples are then pipetted into the remaining wells of the gel. When this is done the lid is placed on the electrophoresis tank making sure that the orientation of the gel and positive and negative electrodes is correct (we want the DNA to migrate across the gel to the positive end).

## Separating the Fragments

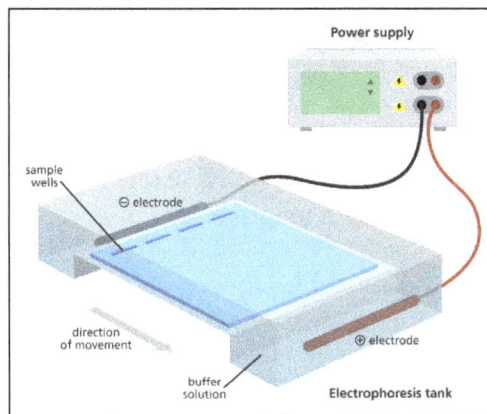

The electrical current is then turned on so that the negatively charged DNA moves through the gel towards the positive side of the gel. Shorter lengths of DNA move faster than longer lengths so move further in the time the current is run.

Illustration of DNA electrophoresis equipment used to separate DNA fragments by size. A gel sits within a tank of buffer. The DNA samples are placed in wells at one end of the gel and an electrical current passed across the gel. The negatively-charged DNA moves towards the positive electrode.

The distance the DNA has migrated in the gel can be judged visually by monitoring the migration of the loading buffer dye. The electrical current is left on long enough to ensure that the DNA fragments move far enough across the gel to separate them, but not so long that they run off the end of the gel.

## Visualising the Results

Once the DNA has migrated far enough across the gel, the electrical current is switched off and the gel is removed from the electrophoresis tank. To visualise the DNA, the gel is stained with a fluorescent dye that binds to the DNA, and is placed on an ultraviolet trans illuminator which will show up the stained DNA as bright bands.

Alternatively the dye can be mixed with the gel before it is poured. If the gel has run correctly the banding pattern of the DNA marker/size standard will be visible. It is then possible to judge the size of the DNA in your sample by imagining a horizontal line running across from the bands of the DNA marker. You can then estimate the size of the DNA in the sample by matching them against the closest band in the marker.

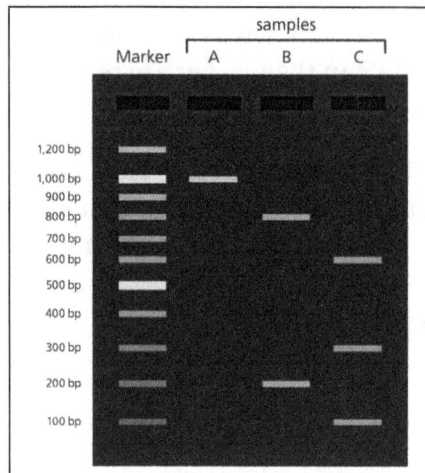

Illustration showing DNA bands separated on a gel. The length of the DNA fragments is compared to a marker containing fragments of known length.

# Blotting

Blotting is technique in which nucleic acids i.e., RNA and DNA or proteins are transferred onto a specific membrane. This membrane may be nitrocellulose PVDF or nylon membrane. This process can be done just after the gel electrophoresis, by transferring the molecules from the gel onto the surface of blotting membrane. But sometimes it can be done by directly transferring the molecules onto the membrane. And then we can visualize these transferring molecules by using staining. Examples: Ethidium bromide, Crystal violet, Safranine and Ossmium tetroxide etc.

## Types of Blotting

## Southern Blotting

Southern blotting is named after Edward M. Southern. This method is used for analysis of DNA sequences. It involves the following steps:

- Firstly, large weighted DNA is cut into small fragments by using Restriction endonucleases.

- Then, these fragments are electrophoresed on separating gel so that they can separate according to their size.

- If DNA fragments are much larger in size so firstly the gel should be treated with HCl, causes depurination of DNA fragments.

- After separating these fragments, placed a nitrocellulose sheet over the separating gel. Apply pressure over the membrane so that proper interaction can occur between these two.

- After that the membrane is exposed to ultraviolet radiation so that the fragments are permanently attached to the membrane.

- Then the membrane is exposed to hybridization probe. But the DNA probe is labeled so that it can easily detect, when the molecule is tagged with a chromogenic dye.

- After hybridization process, excess probe is washed away by using SSC buffer and it can be visualized on X-ray film with the help of autoradiography.

## Applications

- It is used in the technique called RFLP (Restriction fragment length polymorphism) mapping.

- Also used in phylogenetic analysis.

- To identify the gene rearrangements.

## Western Blotting

Western blotting is named after W. Neal Burnette. This method is used for detection and analysis of protein in a given sample. It involves the following steps:

- Firstly, isolating the protein from particular sample.

- After that beta-mercaptoethanol (BME) and Sodium dodecyl Sulfate (SDS) is added into the protein suspension.

- Then, protein-SDS complex is placed on top of the gel in the well. A molecular weight marker is also loaded in one of the well in order to determine the molecular weight of other proteins. After that the samples are added in the remaining wells.

- Once the samples and markers are loaded then current is passed across the gel. Protein is pulled down to the positive pole of the well because it is tightly bound to SDS which is negatively charged. Movement of protein is inversely proportional to its size.

- After this step, gel is placed against a membrane and current is passed across the gel so that all the proteins are transferred onto the membrane.

- Then Immunoblotting has to be done. In this method, firstly block the membrane with non-specific protein in order to prevent antibody from binding to the membrane where the protein is not present.

- After that primary antibody is added to the solution. These antibodies are responsible for recognizing a specific amino-acid sequence. Then wash it to remove unbound primary antibody and add secondary antibody.

- Now these antibodies are conjugated with an enzyme and recognize the primary antibody. Lastly, another wash is done to remove unbound secondary antibody.

- Here, chemiluminescent substrate is used for detection. The light is being emitted once the substrate has been added and can be detected with film imager.

## Applications

- Used in clinical purposes.

- Used to detect specific protein in low quantity.

- Used to quantifying a gene product.

## Northern Blotting

Northern blotting is given by Alwine. This method is used to analyse and detection of RNA in a sample.

- Firstly, extraxt and purify mRNA from the cells.

- Separate these RNA on agarose gels containing formaldehyde as a denaturing agent for the RNA.

- This gel is immersed in depurination buffer for 5-10 minutes and washed with water.

- Then transfer these RNA fragments onto the carrier membrane i.e aminobenzyloxymethyl filter paper.

- After transferring the RNA,it is fixed to the membrane by using UV or heat.

- Add DNA labellled probe for hybridization.

- Wash off the unbound probe and at the end mRNA-DNA hybrid are then detected by X-ray film.

## Applications

- Used in screening.

## Eastern Blotting

Eastern blotting is given by Bogdanov. This method is used to identify carbohydrate epitopes including glycoconjugates and lipids. Mostly blotted proteins after transferring onto the membrane are analyzed for PTMs by using a probe and hence identify carbohydrates and lipids. It involves the following steps:

- Firstly, targeted molecules are vertically separated by using gel electrophoresis.

- Then, these separated molecules are transferred horizontally on the nitrocellulosic membrane.

- After that primary antibody is added to the solution. These antibodies are responsible for recognizing a specific amino-acid sequence. Then wash it to remove unbound primary antibody and add labelled secondary antibody.

- These labelled probes confirm the molecule of interest.

## Applications

- Detection of protein modification.

- Used for binding studies by using various ligands.

- Used to purify various phospholipids.

# DNA Microarray

Nucleic acid arrays or more simply DNA arrays are a group of technologies in which specific DNA sequences are either deposited or synthesized in a 2-D (or sometimes 3-D) array on a surface in such a way that the DNA is covalently or non-covalently attached to the surface. In typical use, a DNA array is used to probe a solution of a mixture of labeled nucleic acids and the binding (by hybridization) of these "targets" to the "probes" on the array is used to measure the relative concentrations of the nucleic acid species in solution. By generalizing to a very large number of spots of DNA, an array can be used to quantify an arbitrarily large number of different nucleic acid sequences in solution. There are other means of quantifying different nucleic acid sequences in a sample, including quantitative PCR, "digital PCR", and hybridization to optically tagged "probe sequences".

One could argue that the original DNA array was created with the colony hybridization method of Grunstein and Hogness. In this procedure, DNA of interest was randomly cloned into *E. coli* plasmids that were plated onto agar petri plates covered with nitrocellulose filters. Replica plating was used to produce additional agar plates. The colonies on the filters were lysed and their DNA's were denatured and fixed to the filter to produce a random and unordered collection of DNA spots that represented the cloned fragments. Hybridization of a radiolabeled probe of a DNA or RNA of interest was used to rapidly screen 1000's of colonies to identify clones containing DNA that was complimentary to the probe.

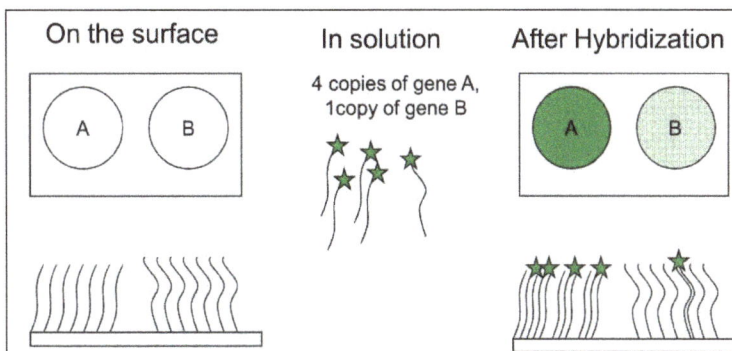

Figure: Simplified view of a DNA array.

The upper rectangles show two spots of DNA on a solid surface (sequences "A" and "B") prior to and after hybridization. The lower rectangles show highly idealized side views of the same surfaces.

In 1979, this approach was adapted to create ordered arrays by Gergen et. al. who picked colonies into 144 well microplates. They created a mechanical 144 pin device and a jig that allowed them to replicate multiple microtiter plates on agar and produce arrays of 1728 different colonies in a

26 × 38 cm region. An additional transfer of colonies to squares of Whatman filter paper followed by a growth, lysis, denaturation and fixing of the DNA to the filter, allowed the production of DNA arrays on filters that could be re-used multiple times. During the next decade, filter based arrays and protocols similar to these were used in a variety of applications including: cloning genes of specific interest, identifying SNP's, cloning genes that are differentially expressed between two samples and physical mapping.

In the late 1980's and early 1990's Hans Lehrach's group automated these processes by using robotic systems to rapidly array clones from microtiter plates onto filter. The concomitant development of cDNA cloning in the late 1970's and early 80's combined with international programs to fully sequence both the human genome and the human transcriptome led to efforts to create reference sets of cDNAs and cDNA filter arrays for human and other genomes By the late 1990's and early 2000's, sets of non-redundant cDNA's became widely available and the complete genome sequences of some organisms allowed for sets of PRC products representing all the known open reading frames (ORFs) in small genomes. These sets, combined with readily available robotics, allowed individual labs to make their own cDNA or ORF arrays that containing gene content that represented the vast majority of genes in a genome.

## Birth of the Modern DNA Array

In the late 90's and 2000's, DNA array technology progressed rapidly as both new methods of production and fluorescent detection were adapted to the task. In addition, increases in our knowledge of the DNA sequences of multiple genomes provided the raw information necessary to assure that arrays could be made which fully represented the genes in a genome, all the sequence in a genome or a large fraction of the sequence variation in a genome. It should also be noted that during this time, there was a gradual transition from spotting relatively long DNA's on arrays to producing arrays using 25-60 bp oligos. The transition to oligo arrays was made possible by the increasing amounts of publicly available DNA sequence information. The use of oligos (as opposed to longer sequences) also provided an increase in specificity for the intended binding target as oligos could be designed to target regions of genes or the genome that were most dissimilar from other genes or regions. Three basic types of arrays came into play during this time frame, spotted arrays on glass, in-situ synthesized arrays and self-assembled arrays.

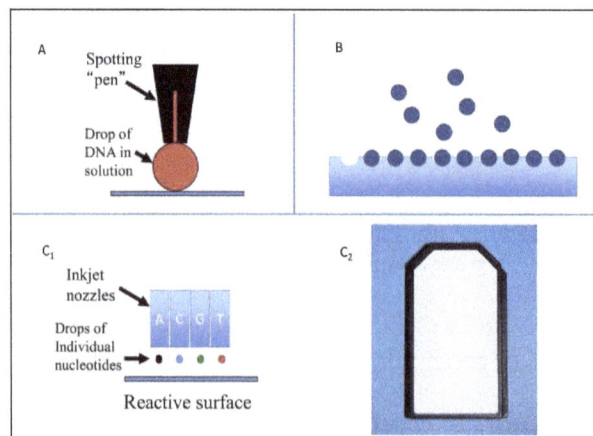

Figure: Three basic types of microarrays: (A) Spotted arrays on glass, (B) self-assembled arrays and (C) in-situ synthesized arrays.

A. With spotted arrays, a "pen" (or multiple pens) are dipped into solutions containing the DNA of interest and physically deposited on a 1 x 3 glass microscope slide. Typically the glass slide surface is coated with something to help retain the DNA such as polylysine, a silane or a chemically reactive surface (to which chemically reactive oligos or PCR products would be added).

B. Self-assembled arrays can be created by applying a collection of beads containing a diverse set of oligos to a surface with pits the size of the beads. After the array is constructed a series of hybridizations determine which oligo is in what position on each unique array.

$C_1$ and $C_2$ - In-situ synthesized arrays can be produced by inkjet oligo synthesis methods ($C_1$) or by photolithographic methods such as used by Affymetrix ($C_2$).

## Spotted Arrays

In 1996 Derisi et. al. published a method which allowed very high-density DNA arrays to be made on glass substrates. Poly-lysine coated glass microscope slides provided good binding of DNA and a robotic spotter was designed to spot multiple glass slide arrays from DNA stored in microtiter dishes. By using slotted pins (similar to fountain pens in design) a single dip of a pin in DNA solution could spot multiple slides. Spotting onto glass, allowed one to fluorescently label the sample. Fluorescent detection provided several advantages relative to the radioactive or chemilluminescent labels common to filter based arrays. First, fluorescent detection is quite sensitive and has a fairly large dynamic range. Second, fluorescent labeling is generally less expensive and less complicated than radioactive or chemilluminescent labeling. Third, fluorescent labeling allowed one to label two (or potentially more) samples in different colors and cohybridize the samples to the same array. As it was very difficult to reproducibly produce spotted arrays, comparisons of individually hybridized samples to ostensibly identical arrays would result in false differences due to array-to-array variation. However, a two-color approach in which the ratio of signals on the same array are measured is much more reproducible.

A method for light directed, spatially addressable chemical synthesis which combined photolabile protecting groups with photolithography to perform chemical synthesis on a solid substrate. In this initial work, the authors demonstrated the production of arrays of 10-amino acid peptides and, separately, arrays of di-nucleotides. In 1994, Fodor et.al. at the recently formed company of Affymetrix demonstrated the ability to use this technology to generate DNA arrays consisting of 256 different octa-nucleotides. By 1995-1996, Affymetrix arrays were being used to detect mutations in the reverse transcriptase and protease genes of the highly polymorphic HIV-1 genome and to measure variation in the human mitochondrial genome. Eventually, Affymetrix used this technology to develop a wide catalogue of DNA arrays for use in expression analysis, genotyping and sequencing for the current catalog of arrays.

A major advantage of the Affymetrix technology is that because the DNA sequences are directly synthesized on the surface, only a small collection of reagents (the 4 modified nucleotides, plus a small handful of reagents necessary for the de-blocking and coupling steps) are needed to construct an arbitrarily complex array. This contrasts with the spotted array technologies in which one needed to construct or obtain all the sequences that one wished to deposit on the array in advance of array construction. However, the initial Affymetrix technology was limited in flexibility as each model of array required the construction of a unique set of photolithographic masks in order to direct the light to the array at each step of the synthesis process. In 2002, authors from Nimblegen Systems Inc., published a method in which the photo-deprotection step of Fodor et. al is accomplished using micro-mirrors (similar to those in video computer projectors) to direct light at the pixels on the array. This allows for custom arrays to be manufactured in small volumes at much lower cost than by photolithographic methods using masks to direct light (which are cheaper for large volume production). One constraint with this method is that the total number of addressable pixels (e.g. unit oligos that can be synthesized) is limited to the number of addressable positions in the micro-mirror device (of order 1M).

In 1996, Blanchard et.al. proposed a method use inkjet printing technology and standard oligo synthesis chemistry to produce oligo arrays. In brief, inkjet printer heads were adapted to deliver to the four different nucleotide phosphoramidites to a glass slide that was pre-patterned to contain regions containing hydrophilic regions (with exposed hydroxyl groups) surrounded by hydrophobic regions. The hydroxylated regions provided a surface to which the phosphoramidites could couple, while the surrounding hydrophobic regions contained the droplet(s) emitted by the inkjets to defined regions. This technology was eventually commercialized by Rosetta Inpharmatics and licensed to Agilent Technologies who produces these arrays at present. The inkjet array approach shares the advantage of the Affymetrix/Nimblegen approach in that one only need to have available a small number of reagents to produce an array. In addition, similar to the Nimblegen approach, the production of a new type of array only requires that a different set of sequence information is delivered to the printer. Hence, the inkjet array technology has been particularly useful for the design of custom arrays that are produced in low volume.

## Self-assembled Arrays

An alternative approach to the construction of arrays was created by the group of David Walt at Tufts University and ultimately licensed to Illumina. Their method involved synthesizing DNA on small polystryrene beads and depositing those beads on the end of a fiber optic array in which the ends of the fibers were etched to provide a well that is slightly larger than one bead. Different types of DNA would be synthesized on different beads and applying a mixture of beads to the fiber optic cable would result in a randomly assembled array. In early versions of these arrays, the beads were optically encoded with different fluorophore combinations in order to allow one to determine which oligo was in which position on the array (referred to as "decoding the array"). Optical decoding by fluorescent labeling limited the total number of unique beads that could be distinguished. Hence, the later and present day methods for decoding the beads involve hybridizing and detecting a number of short, fluorescently labeled oligos in a sequential series of steps. This not only allows for an extremely large number of different types of beads to be used on a single array but also functionally tests the array prior to its use in a biological assay. Later versions of the Illumina arrays used a pitted glass surface to contain the beads instead of a fiber option arrays.

## Applications of Microarrays

### Gene Expression Analysis

The predominate application of DNA microarrays has been to measure gene expression levels. In this application, RNA is extracted from the cells of interest and either, labeled directly, converted to a labeled cDNA or converted to a T7 RNA promoter tailed cDNA which is further converted to cRNA through the Eberwine amplification process. A wide variety of methods have been developed for labeling of the cDNA or cRNA including: incorporation of fluorescently labeled nucleotides during the synthesis, incorporation of biotin labeled nucleotide which is subsequently stained fluorescently labeled streptavidin, incorporation of a modified reactive nucleotide to which a fluorescent tag is added later, and a variety of signal amplification. The two most frequently used methods are the incorporation of fluorescently labeled nucleotides in the cRNA or cDNA synthesis step or the incorporation of a biotin labeled nucleotide in the cRNA synthesis step (as is done by Affymetrix).

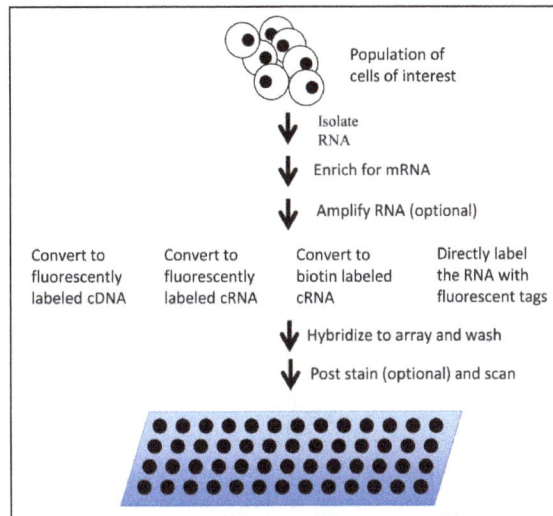

Figure: Gene expression analysis via microarrays.

RNA is isolated from the sample of interest and enriched for messenger RNA. In eucaryotes, poly-A tailed mRNA's are typically enriched using affinity purification with oligo dT beads or columns. In procaryotes, unselected RNA is typically depleted for ribosomal sequences using bead or columns coated with sequences complementary to 16s. After message enriched RNA is in hand, it is optionally amplified and labeled by any one of a number of methods and the resulting labeled sample is hybridized to a microarray. The array is washed to remove unbound sample. If the sample was labeled with biotin, the array is post stained with fluorescently labeled streptavidin and washed again. The array is then scanned to measure fluorescence signal at each spot on the array.

The labeled cRNA or cDNA are then hybridized to the microarray, the array is washed and the signal is detected by measuring fluorescence at each spot. In the case of biotin labeled samples, the array is stained post-hybridization with fluorescently labeled streptavidin. Laser induced fluorescence is typically measured with a scanning confocal microscope. The intensity of the signal(s) on each spot is taken as a measure of the expression level of the corresponding gene.

## Transcription Factor Binding Analysis

Microarrays have also been used in combination with chromatin immunoprecipitation to determine the binding sites of transcription factors. In brief, transcription factors (TFs) are cross linked to DNA with formaldehyde and the DNA is fragmented. The TF(s) of interest (with the DNA to which they were boud still attached) are affinity purified using either an antibody to the TF or by tagging the transcription factor with peptide that's amenable to affinity chromatography (for example a FLAG-, HIS-, myc or HA-tag). After purification, the DNA is released from the TF, amplified, labeled and hybridized to the array. This technique is commonly referred to as "ChIP-chip" for Chromatin Immuno-Precipitation on a "chip" or microarray.

As TF's often bind quite a distance away from the genes that they regulate, the design of the array and size distribution of the fragment length are interrelated. E.g. the array must contain probes that will interrogate the region of DNA bound to the transcription factor. For bacteria or yeast, the intergenic regions are fairly small and the same arrays used for gene expression work can be applied to ChIP-chip. For mammalian genomes, the intergenic regions are large and the TF often bind many kbp away from the gene of interest. Hence, for mammalian genomes, oligo arrays with oligo's spaced evenly across the entire genome are typically used for ChIP-chip experiments. Buck et. al. provide a good review of the considerations for the design and analysis of ChIP-chip experiments.

## Genotyping

Microarrays have been widely used as single-nucleotide-polymorphism (SNP) genotyping platforms. Several alternative approaches have been used to detect SNP's but the most commonly used are allele discrimination by hybridization as used by Affymetrix, allele specific extension and ligation to a "bar-code" oligo which is hybridized to a universal array (the Illumina "Golden Gate Assay") or approaches in which the arrayed DNA is extended across the SNP in a single nucleotide extension reaction. explains the detection approaches in more detail. Allelic discrimination by hybridization suffers background due to non-specific hybridization in complex genomes. In order to reduce this background, Affymetrix developed a PCR based approach to reduce genomic complexity.

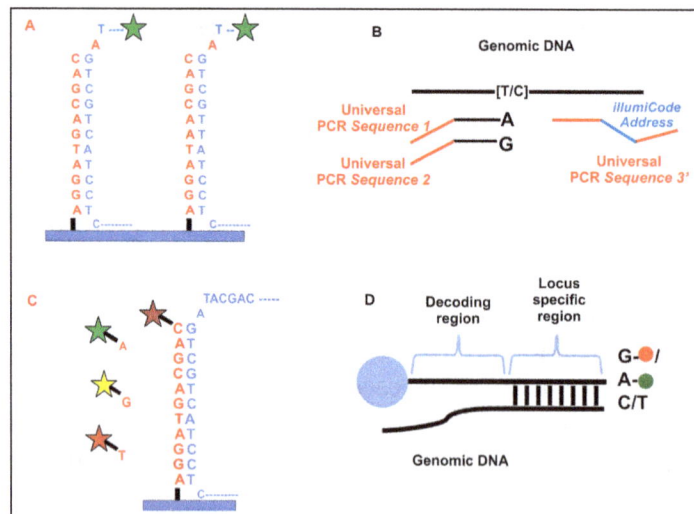

Figure: SNP detections strategies for arrays.

A) Allele discrimination by hybridization – Oligos that are complimentary to each allele are placed on the array and labeled genomic DNA is hybridized to the array. The variant position is placed in the center of the oligo (typically 25 bp on Affymetrix arrays) as this position has the greatest affect on hybridization. Typically, multiple array positions are used for each allele to improve signal to noise. B) Illumina's "Golden Gate Assay"- two allele specific oligos are each tailed with a different universal primer (1 and 2) and hybridized in solution to genomic DNA. A third oligo that is complementary to the same locus is tailed with a "barcode" sequence and a third universal primer (3). Polymerase is used to extend the allele specific primers across the genomic sequence and the extended products are ligated to the third oligo. PCR is performed using primers complimentary to universal sequences 1, 2 and 3. The PCR primers complimentary to the universal sequences 1 and 2 are labeled with a unique fluorophore. The barcode sequence on the third oligo allows the PCR product to be uniquely detected on an array containing oligos complimentary to the barcode sequence. The use of multiple barcodes (one for each locus of interest) allows the assay to be multiplexed to sample many loci. C) Arrayed primer extension (APEX) – In this assay, the array contains DNA oriented with the 5' end attached to the array and the 3' end stopping one nucleotide short of the SNP. Genomic DNA is fragmented and hybridized to the array and the oligo on the array is extended in single nucleotide dye terminator sequencing reaction. D) Illumina's Infinium assay – This assay is similar to the APEX assay except that the oligo to be extended is on a bead and the single nucleotide that is added is labeled with a nucleotide specific hapten as opposed to a fluorophore. The haptens are then detected by staining with fluorescently labeled proteins that bind each hapten.

In brief, SNPs for their assay are selected to be between restriction sites that are <1kb apart. Genomic DNA is fragmented with a restriction enzyme, end repaired and adapters for PCR are ligated to the fragments. PCR is performed under conditions that selectively amplify products of <1kb in size. This method reduces genomic complexity by approximately 50-fold and results in a corresponding increase in signal to noise on the array.

Both the Affymetrix and the Illumina methods for SNP genotyping have been highly successful and are highly used. Today SNP arrays capable of detecting >1M different human SNPs are available from both vendors. Call rates and reproducibility of SNP calls exceed 99.5%. In addition, the same arrays or variations thereof can also be used to detect copy number variants.

## Data Standards and Data Exchange

With the exception of DNA sequencing, microarrays were perhaps the earliest technology that allowed biologists to vast amounts of complex digital data. As the technology came into use, it rapidly became apparent that in order for others to be able to reproduce a given microarray experiment a detailed description of the array, the sample, the protocols and the data analysis methods needed to be available. Moreover, it also became apparent that access to the raw and processed data would allow others to perform analyses and Meta analyses (on combinations of data) that the original data producers had not conceived. To address these issues of reproducible science and data exchange, members of the Microarray Gene Expression Data Society (now the Function Genomics Data Society) created the MIAME (Minimum Information About a Microarray Experiment) standards for the description of microarray experiments and for the exchange of microarray data. These efforts influenced the creation public databases for microarray data and subsequent standards efforts in other areas.

## Limitations of DNA Microarrays

At their core, microarrays are simply devices to simultaneously measure the relative concentrations of many different DNA or RNA sequences. While they have been incredibly useful in a wide variety of applications, they have a number of limitations. First, arrays provide an indirect measure of relative concentration. That is the signal measured at a given position on a microarray is typically assumed to be proportional to the concentration of a presumed single species in solution that can hybridize to that location. However, due to the kinetics of hybridization, the signal level at a given location on the array is not linearly proportional to concentration of the species hybridizing to the array. At high concentrations the array will become saturated and at low concentrations, equilibrium favors no binding. Hence, the signal is linear only over a limited range of concentrations in solution. Second, especially for complex mammalian genomes, it is often difficult (if not impossible) to design arrays in which multiple related DNA/RNA sequences will not bind to the same probe on the array. A sequence on an array that was designed to detect "gene A", may also detect "genes B, C and D" if those genes have significant sequence homology to gene A. This can particularly problematic for gene families and for genes with multiple splice variants. It should be noted that it is possible to design arrays specifically to detect splice variants either by making array probes to each exon in the genome or to exon junctions. However, it is difficult to design arrays that will uniquely detect every exon or gene in genomes with multiple related genes.

Finally, a DNA array can only detect sequences that the array was designed to detect. That is, if the solution being hybridized to the array contains RNA or DNA species for which there is no complimentary sequence on the array, those species will not be detected. For gene expression analysis, this typically means that genes that have not yet been annotated in a genome will not be represented on the array. In addition, non-coding RNA's that are not yet recognized as expressed are typically not represented on an array. Moreover, for highly variable genomes such as those from bacteria, arrays are typically designed using information from the genome of a reference strain. Such arrays may be missing a large fraction of the genes present in a given isolate of the same species. For example, in the bacterial species Aggregatibacter actinomycetemcomitans, the gene content differs by as much as 20% between any two isolates. Hence an array designed using gene annotation from a "reference isolate" will not contain many of the genes found in other isolates.

## Future of DNA arrays

Given the limitations of arrays mentioned above, it would be far preferable to have an unbiased method to directly measure all the DNA or RNA species present in a particular sample. The advent of next generation sequencing technologies combined with the rapid decrease in the cost of sequencing has now made sequencing cost competitive with microarrays for all assays with the possible exception of genotyping. When the cost is similar, sequencing has many advantages relative to microarrays. Sequencing is a direct measurement of which nucleic acids are present in solution. One need only count the number of a given type of sequences present to determine its abundance. Counting sequences is linear with concentration and the signal to noise one can obtain by sequencing is only limited by the number of reads used for each sample.

Sequencing is a relatively unbiased approach to measuring which nucleic acids are present in solution. While sample preparation or different enzymes may bias sequencing counts, unlike DNA arrays, sequencing is not dependent on prior knowledge of which nucleic acids may be present. Sequencing is also able to independently detect closely related gene sequences, novel splice forms or RNA editing that may be missed due to cross hybridization on DNA microarrays. As a result of these advantages and the decreasing cost of sequencing, DNA arrays are being rapidly replaced by sequencing for nearly every assay that has been previously performed on microarrays. As the cost of sequencing is currently dropping by a factor of two every five months, it's likely that DNA arrays will be fully replaced by sequencing methods within the next 5-10 years.

# Allele-specific Oligonucleotide

Although unknown mutations can be investigated in some cases, most of the molecular testing is performed in order to find known gene variants, like mutational hot spots, founder mutations, molecular risk factors or mutations/polymorphisms of pharmacogenetic importance. Traditionally, the most widely used method for the detection of point mutations is the PCR followed by restriction digestion (PCR-RFLP, restriction fragment length polymorphism). As the bacterial restriction endonucleases recognize a very specific sequence motif, they are very useful fotr the detection of mutations affecting restriction sites. There is a few hundred such enzyme. During the testing, the last step is an electrophoresis on agarose or acrylamide to separate the fragments according to their size. The number and size distribution of the restriction fragments are the function of the genotype.

Allele-specific PCR method can be used, when the site of the mutation is known, as a mutation detection method. Allele-specific PCR is based on the fact, that Taq polymerase enzyme needs perfectly hybridized primer (on its 3' end) for its function. In the case of 3' incomplementarity, Taq enzyme is unable to use the oligonucleotide as primer. The system contains a common oligonucleotide and contains primers specific to the possible alleles. Allele-specific PCR reactions are separated physically. With careful optimization, allele-specific amplification is possible. In the upper part of the picture a perfect hybridization is shown resulting in PCR product. In the lower panel, no product is formed due to the 3' end incomplementarity. Using several pairs of primers, multiplexing is possible. The method is capable of showing all possible genotypes, i.e., wild type, heterozygous, homozygous.

Allele-specific PCR has been popular under many names, like ASO-PCR, ARMS. There are commercially available mutation detection diagnostic kits that are based on this principle.

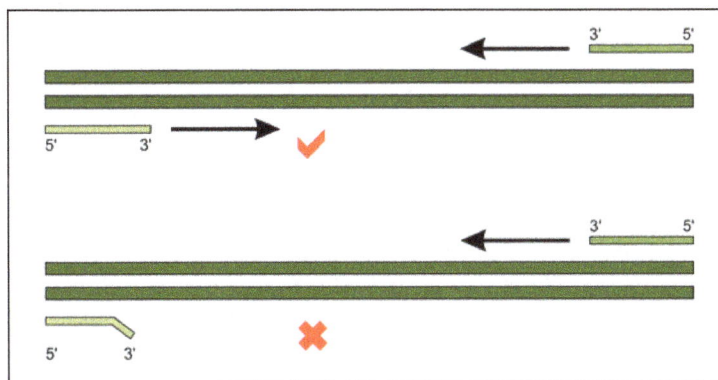

Figure: Allele-specific PCR.

Although the allele specific PCR has significant advantages compared to PCR-RFLP as it can be multiplexed, it still requires electrophoretic separation. There have been many efforts to develop a methodology that does not involve electrophoresis during allele discrimination.

Allele specific oligonucleotide hybridization is a general method distinguishing between alleles that differ by even a single nucleotide substitution. The method is based on the difference in the melting temperature between perfectly matched and mismatched probe-template hybrids. In this procedure the allele specific oligonucleotides (ASOs) are immobilized on a nitrocellulose or nylon membrane and hybridized with a labelled PCR product spanning the variant nucleotide site. The discrimination between the two alleles is based on the fact that in a specific hybridization temperature the perfectly matched hybrid is more stable than the mismatched one. After hybridization and washing the detection is mainly colorimetric: for example if the PCR product is labelled with biotin, the detection is based on the streptavidin- alkaline phosphatase conjugate (or horseradish peroxidase) enzyme reaction with a substrate. By using two ASOs, it is possible to determine each combination: wild type, heterozygous, homozygous. The advantage of this method that multiplexing is possible (at least 5-10 different mutations are detectable at the same time) and it does not involve a gel electrophoretic step, though many washing steps are needed and high stringency conditions are required. There are commercially available mutation detection diagnostic kits that are based on this principle.

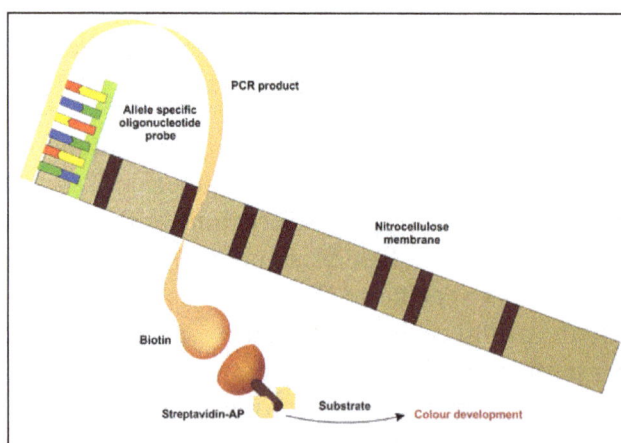

Figure: Mutation detection using allele-specific oligonucleotide hybridization.

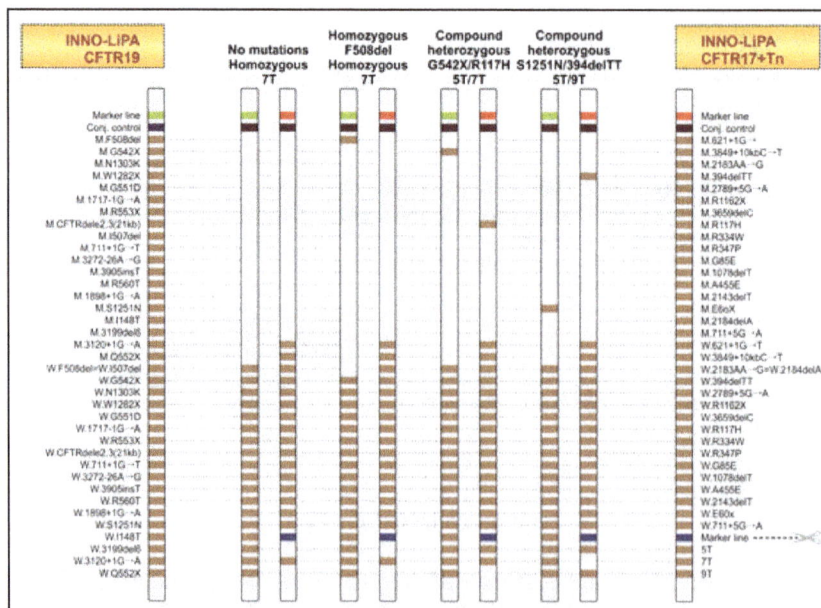

Figure: Mutation detection using allele-specific oligonucleotide hybridization for the most common mutations causing cystic fibrosis.

One of the most important use of the allele specific oligonucleotide hybridization is shown in figure above. Mutations in the CFTR gene which cause cystic fibrosis (CF) are very heterogeneous. To date, more than 1600 CF-causing mutations have been described of which approximately 30 are more common. There are mutation testing panels to analyze the most common alterations and the available diagnostic kits include these mutations. The figure shows such an assay. The test system, which is highly multiplexed, is performed using two membane strips. Mutations are represented both with their wild type and mutated alleles. The left-hand side of the picture shows the CFTR19 and right side shows the CFTR17+Tn kit. Using both strips, many different mutations can be tested simultaneously. Sample 1 has been shown to be wild type for the analyzed mutations while sample 2 is homozygous mutant for p.F508del, the far most common CF-causing mutation as judged by the respective signals: on the CFTR19 strip only the M.F508del allele-specific probe provides signal, the W.F508del = W.I507del probe does not. Sample 3 has two mutations. M.G542X could be detected on the CFTR19 strip, and M.R117H gives a signal on the CFTR17+Tn strip. As in both cases the wild type probe also gave a signal, the genotype of the sample is compound heterozygous. Sample 4 was similar to sample 3. The detected alterations were p.S1251N and 394delTT. Using this – or similar – diagnostic approach, 80-95% of the pathogenic mutations causing CF can be detected, depending on the tested population.

## References

- Foundations-of-molecular-cloning-past-present-and-future, tools-and-resources: international.neb.com, Retrieved 1 July, 2019

- Polymerase-chain-reaction-pcr, dna-sequencing-pcr-electrophoresis, biotech-dna-technology, biology, science: khanacademy.org, Retrieved 20 February, 2019

- What-is-gel-electrophoresis: yourgenome.org, Retrieved 23 August, 2019

- Types-of-blotting: rroij.com, Retrieved 13 March, 2019

- Molelkularis-diagnoszitka: tankonyvtar.hu, Retrieved 18 January, 2019

# Chapter 4

# DNA and Genome: Organization and Sequencing

The linear order of DNA elements as well as their division into chromosomes is referred to as genomic organization. DNA sequencing refers to the process which is used to determine the nucleic acid sequence. This chapter has been carefully written to provide an easy understanding of these processes related to the organization and sequencing of genomes and DNA as well as the technologies used for them.

## Structure and Function of DNA

### Molecular Structure

- Primary structure of the molecule-covalent backbone and bases aside. A nucleoside is made of a sugar + a nitrogenous base.

- A nucleotide is made of a phosphate + a sugar + a nitrogenous base. In DNA, the nucleotide is a deoxyribonucleotide (in RNA, the nucleotide is a ribonucleotide).

### Phosphoric Acid

- Gives a phosphate group.

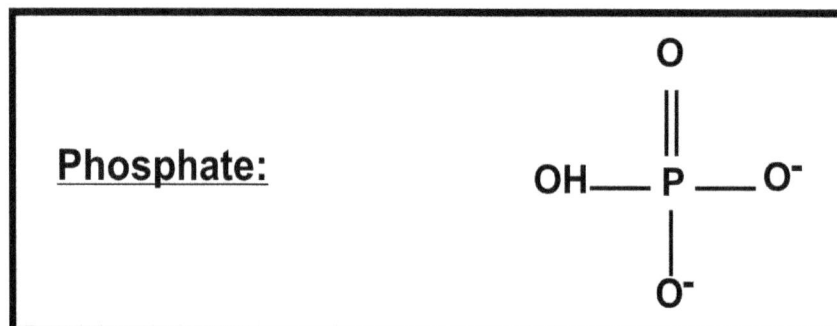

### Sugar

Deoxyribose, which is a cyclic pentose (5-carbon sugar). Note: the sugar in RNA is a ribose. Carbons in the sugar are noted from 1' to 5'. A nitrogen atom from the nitrogenous base links to C1' (glycosidic link), and the phosphate links to C5' (ester link) to make the nucleotide. The nucleotide is therefore: phosphate - C5' sugar C1' - base.

**Deoxyribose** / **Ribose**

## Nitrogenous Bases

Aromatic heterocycles; there are purines and pyrimidines.

- Purines: adenine (A) and guanine (G).

- Pyrimidines: cytosine (C) and thymine (T) (Note: thymine is replaced by uracyle (U) in RNA).

Other nitrogenous bases exist, in particular methylated bases; methylation of the bases has a functional role.

Pyrimidine / Uracil (U) RNA only / Thymine (T) DNA only / Cytosine (C) both DNA and RNA

Purine / Adenine (A) / Guanine (G)

- Nucleoside names: deoxyribonucleosides in DNA: deoxyadenosine, deoxyguanosine, deoxycytidine, deoxythymidine in DNA (ribonucleosides in RNA: adenosine, guanosine, cytidine, uridine).

- Nucleotide names: deoxyribonucleotides in DNA: deoxyadenylic acid, deoxyguanylic acid, deoxycytidylic acid, deoxythymidylic acid (ribonucleotides in RNA: adenylic acid, guanylic acid, cytidylic acid, uridylic acid).

Figure showing nitrogenous bases, base pairs, sugar-phosphate backbone, and nucleotide structure labeled (a), (b), and (c).

## Three-dimentional Conformation of DNA

## Dinucleotides

Dinucleotides form from a phosphodiester link between 2 mononucleotides. The phosphate of a mononucleotide (in C5′ of its sugar) being linked to the C 3′ of the sugar of the previous mononucleotide. Then, we start with a phosphate, a 5′ sugar (+base) and the 3′ of this sugar, linked to a second phosphate - 5′ sugar, which 3′ is free for next step. The link and the orientation of the molecule is therefore 5′ -> 3′. Polynucleotides are made of the successive addition of monomeres in a general 5′ -> 3′ configuration. The backbone of the molecule is made of a succession of phosphate-sugar (nucleotide n) - phosphate-sugar (nucleotide n+1), and so on, covalently linked, the bases being aside.

## DNA Molecule

DNA is made of two ("duplex DNA") dextrogyre (like a screw; right-handed) helical chains or strands ("the double helix"), coiled around an axis to form a double helix of 20 Å of diameter. The two strands are antiparallel (id est: their 5'->3' orientations are in opposite direction). The general appearance of the polymere shows a periodicity of 3.4 Å, corresponding to the distance between 2 bases, and another one of 34 Å, corresponding to one helix turn (and also to 10 bases pairs).

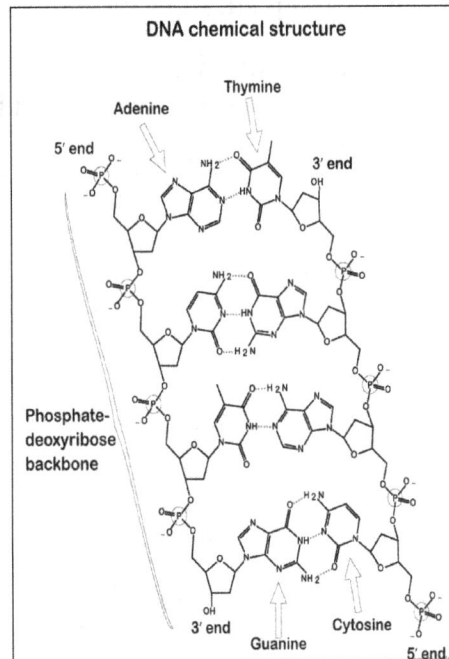

## Hydrogen Bounds: Bases Pairing

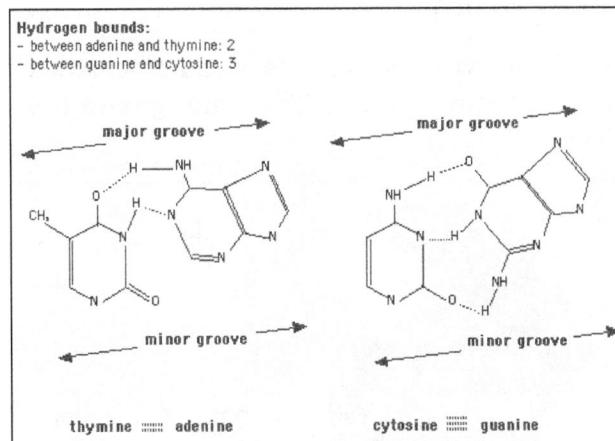

The (hydrophobic) bases are stacked on the inside, there planes are perpendicular to the axis of the double helix. The outside (phosphate and sugar) are hydrophilic. Hydrogen bounds between the bases of one strand and that of the other strand hold the two strands together (dashed lines in the drawing). A purine on one strand shall link to a pyrimidine on the other strand. As a corollary, the number of purines residues equals the number of pyrimidine residues.

- A binds T (with 2 hydrogen bounds).

- G binds C (with 3 hydrogen bounds: more stable link: 5.5 kcal vs 3.5 kcal).

The content in A in the DNA is therefore equal to the content in T, and the content in G equals the content in C. This strict correspondence (A<->T and G<->C) makes the 2 strands complementary. One is the template of the other one, and reciprocally: this property will allow exact replication (semi-conservative replication: one strand and the template is conserved, another is newly synthesized, same with the second strand, conserved, allowing another one to be newly synthesized).

Hydrogen bounds in base pairing are sometimes different from the model of Watson and Crick above described, using the N7 atom of the purine instead of the N1 (Hoogsteen model).

## Major Groove and Minor Groove

The double helix is a quite rigid and viscous molecule of an immense length and a small diameter. It presents a major groove and a minor groove. The major groove is deep and wide, the minor groove is narrow and shallow.

DNA-protein interactions are major/essential processes in the cell life (transcription activation or repression, DNA replication and repair). Proteins bind at the floor of the DNA grooves, using specific binding to hydrogen bounds and nonspecific binding: van der Waals interactions, generalized electrostatic interactions; proteins recognize H-bond donnors, H-bond acceptors, metyl groups (hydrophobic), the latter being exclusively in the major groove; there are 4 possible patterns of recognition with the major groove, and only 2 with the minor groove.

- Some proteins bind DNA in its major groove, some other in the minor groove, and some need to bind to both.

- The 2 strands are called "plus" and "minus" strands, or "direct" and "reverse" strands. At a given location where one strand (any of the two) bears coding sequences, it is unlikely (but not impossible) that the other strand also bears coding sequences.

- DNA is ionized in vivo and behave like a polyanion.

- The double helix as described above is the "B" form of the DNA; it is the form the most commonly found in vivo, but other forms exist in vivo or in vitro. The "A" form resemble B-DNA but it is less hydrated than B-DNA, "A" form is not found in vivo.

## Non B-DNA

DNA is a molecule which moves, fidgets, does gymnastics, dances. The structures are being proved to have funtional roles; on the other hand, they may favour DNA breaks and further deletions, amplification, recombination, and mutations.

## Z-DNA

Z form is a levogyre (left handed) double helix with a zig-zag conformation of the backbone (less smooth than B-DNA). Only one groove is observed, resembling the minor groove, the base pairs being set off to the side, far from the axis. The bases (which form the major groove -close to the axis- in B-DNA) are here at the outer surface. Phosphates are closer together than in B-DNA. Z-DNA cannot form nucleosomes.

A high G-C content favours Z conformation. Cytosine methylation and molecules which can be present in vivo such as spermine and spermidine can stabilize Z conformation.

- DNA sequences can flip from a B form to a Z form and vice versa: Z-DNA is a transient form in vivo.

- Z-DNA formation occurs during transcription of genes, at transription start sites near promoters of actively transcribed genes. During transcription, the movement of RNA polymerase induces negative supercoiling upstream and positive supercoiling downstream the site of transcription The negative supercoiling upstream favours Z-DNA formation; a Z-DNA function would be to absorb negative supercoiling. At the end of transcription, topoisomerase relaxes DNA back to B conformation.

- Certain proteins bind to Z-DNA, in particular double-stranded RNA adenosine deaminase

(ADAR1), a Z-DNA binding nuclear-RNA-editing enzyme; this enzyme converts adenine to inosine in the pre-mRNA. Following, ribosomes will interpret inosine as guanine, and the protein coded with this epigenetic modification will be different.

- Z-DNA antibodies are found in lupus erythematosus and other autoimmune diseases.

- Double stranded RNA (dsRNA) can adopt a Z conformation.

## Cruciform DNA and Hairpin DNA

Holliday junctions (formed during recombination) are cruciform structures. Inverted (or mirror) repeats (palindromes) of polypurine/polypyrimidine DNA stretches can also form cruciform or hairpin structures through intra-strand pairing. Palindromic AT-rich repeats are found at the breakpoints. The only known recurrent constitutional reciprocal translocation. Nucleases bind and cleave holiday junctions after recombination. Other well-known proteins such as HMG proteins and MLL can also bind cruciform DNA.

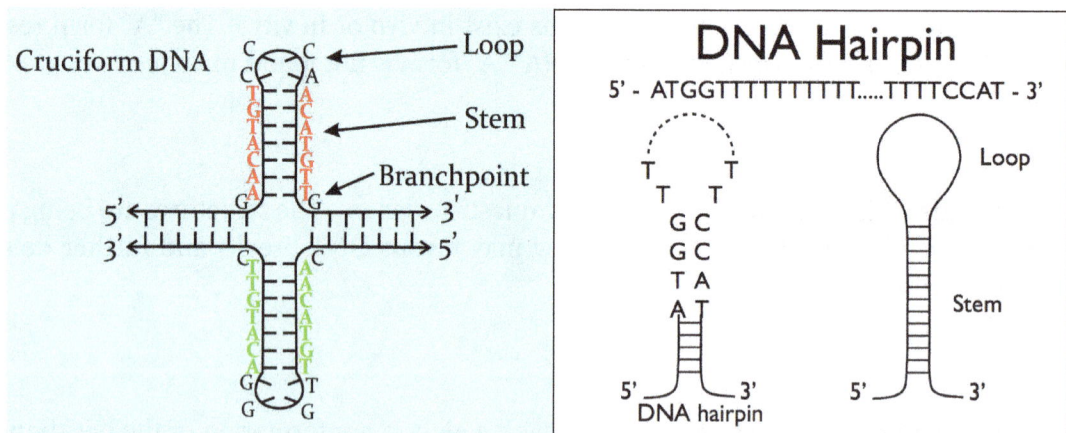

## H-DNA or Triplex DNA

Inverted repeats (palindromes) of polypurine/polypyrimidine DNA stretches can form triplex structures (triple helix). A triple-stranded plus a single stranded DNA are formed. H-DNA may have a role in functional regulation of gene expression as well as on RNAs (e.g. repression of transcription).

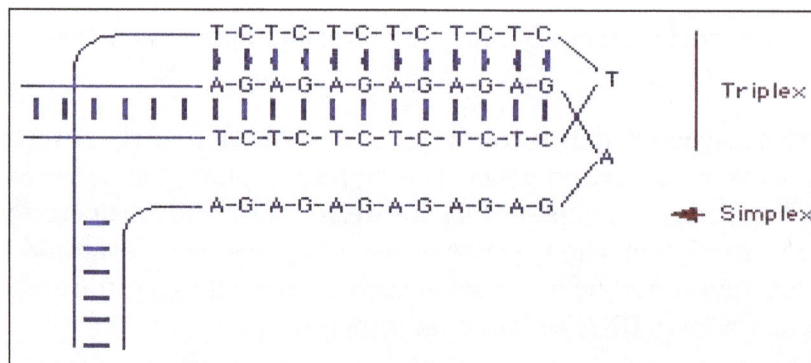

Triplex DNA

# G4-DNA

G4 DNA or quadruplex DNA folding of double stranded GC-rich sequence onto itself forming Hoogsteen base pairing between 4 guanines ("G4"), a highly stable structure. Often found near promotors of genes and at the telomeres. Role in meiosis and recombination; may be regulatory elements. RecQ family helicases are able to unwind G4 DNA (e.g. BLM, the gene mutated in Bloom syndrome.

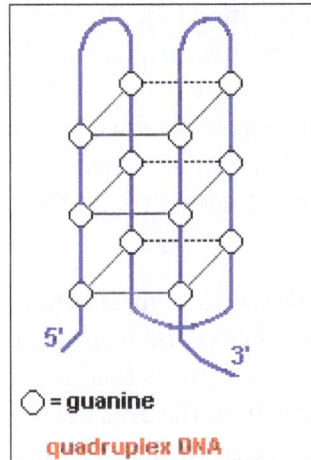

## Quaternary Structure of the Molecule - Chromatin

DNA is associated with proteins: histones and non-histone proteins, to form the chromatin. DNA as a whole is acidic (negatively charged) and binds to basic (positively charged) proteins called histones. There is $3 \times 10^9$ nucleotide pairs in the human haploid genome representing about 30 000 genes dispersed over 23 chromosomes for a haploid set.

## DNA Molecule Consists of Two Complementary Chains of Nucleotides

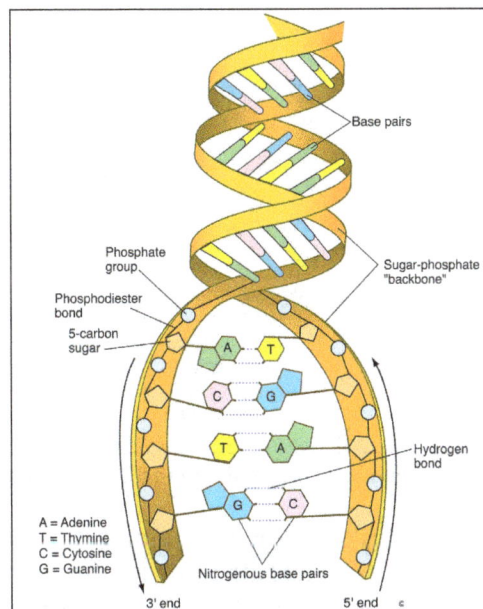

DNA and its building blocks.

A DNA molecule consists of two long polynucleotide chains composed of four types of nucleotide subunits. Each of these chains is known as a DNA chain, or a DNA strand. Hydrogen bonds between the base portions of the nucleotides hold the two chains together. The nucleotides are composed of a five-carbon sugar to which are attached one or more phosphate groups and a nitrogen-containing base. In the case of the nucleotides in DNA, the sugar is deoxyribose attached to a single phosphate group (hence the name deoxyribonucleic acid), and the base may be either adenine (A), cytosine (C), guanine (G), or thymine (T). The nucleotides are covalently linked together in a chain through the sugars and phosphates, which thus form a "backbone" of alternating sugar-phosphate-sugar-phosphate. Because only the base differs in each of the four types of subunits, each polynucleotide chain in DNA is analogous to a necklace (the backbone) strung with four types of beads (the four bases A, C, G, and T). These same symbols (A, C, G, and T) are also commonly used to denote the four different nucleotides—that is, the bases with their attached sugar and phosphate groups.

DNA is made of four types of nucleotides, which are linked covalently into a polynucleotide chain (a DNA strand) with a sugar-phosphate backbone from which the bases (A, C, G, and T) extend. A DNA molecule is composed of two DNA strands held together by hydrogen bonds between the paired bases. The arrowheads at the ends of the DNA strands indicate the polarities of the two strands, which run antiparallel to each other in the DNA molecule. In the diagram at the bottom left of the figure, the DNA molecule is shown straightened out; in reality, it is twisted into a double helix, as shown on the right.

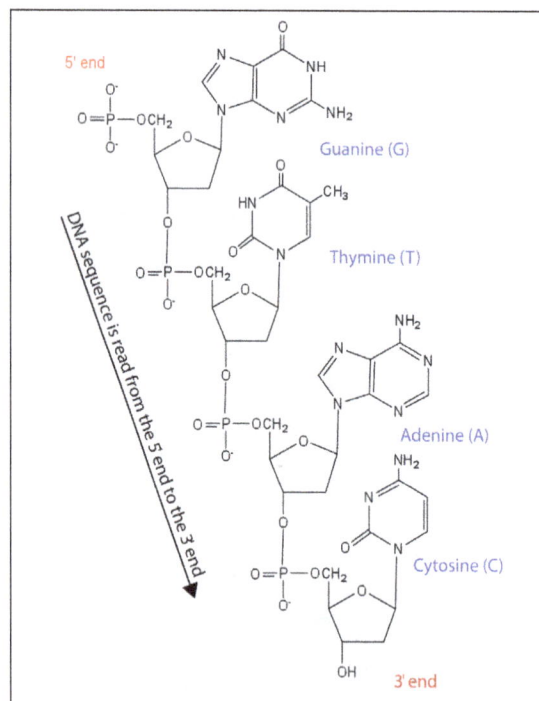

Figure: Complementary base pairs in the DNA double helix.

The way in which the nucleotide subunits are lined together gives a DNA strand a chemical polarity. If we think of each sugar as a block with a protruding knob (the 5′ phosphate) on one side and a hole (the 3′ hydroxyl) on the other, each completed chain, formed by interlocking knobs with holes, will have all of its subunits lined up in the same orientation. Moreover, the two ends of the chain will be easily distinguishable, as one has a hole (the 3′ hydroxyl) and the other a knob (the 5′ phosphate) at its terminus. This polarity in a DNA chain is indicated by referring to one end as the 3′ end and the other as the 5′ end.

The three-dimensional structure of DNA—the double helix—arises from the chemical and structural features of its two polynucleotide chains. Because these two chains are held together by hydrogen bonding between the bases on the different strands, all the bases are on the inside of the double helix, and the sugar-phosphate backbones are on the outside. In each case, a bulkier two-ring base is paired with a single-ring base (a pyrimidine); A always pairs with T, and G with C. This complementary base-pairing enables the base pairs to be packed in the energetically most favorable arrangement in the interior of the double helix. In this arrangement, each base pair is of similar width, thus holding the sugar-phosphate backbones an equal distance apart along the DNA molecule. To maximize the efficiency of base-pair packing, the two sugar-phosphate backbones wind around each other to form a double helix, with one complete turn every ten base pairs.

The shapes and chemical structure of the bases allow hydrogen bonds to form efficiently only between A and T and between G and C, where atoms that are able to form hydrogen bonds can be brought close together without distorting the double helix. As indicated, two hydrogen bonds form between A and T, while three form between G and C. The bases can pair in this way only if the two polynucleotide chains that contain them are antiparallel to each other.

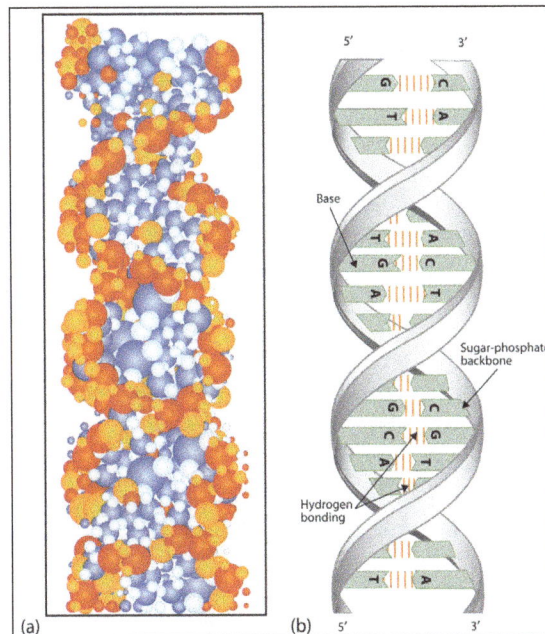

Figure: The DNA double helix.

(A) A space-filling model of 1.5 turns of the DNA double helix. Each turn of DNA is made up of 10.4 nucleotide pairs and the center-to-center distance between adjacent nucleotide pairs is 3.4 nm. The coiling of the two strands around each other creates two grooves in the double helix. As indicated in the figure, the wider groove is called the major groove, and the smaller the minor groove. (B) A short section of the double helix viewed from its side, showing four base pairs. The nucleotides are linked together covalently by phosphodiester bonds through the 3′-hydroxyl (-OH) group of one sugar and the 5′-phosphate (P) of the next. Thus, each polynucleotide strand has a chemical polarity; that is, its two ends are chemically different. The 3' end carries an unlinked -OH group attached to the 3' position on the sugar ring; the 5' end carries a free phosphate group attached to the 5' position on the sugar ring.

The members of each base pair can fit together within the double helix only if the two strands of the helix are antiparallel—that is, only if the polarity of one strand is oriented opposite to that of the other strand. A consequence of these base-pairing requirements is that each strand of a DNA molecule contains a sequence of nucleotides that is exactly complementary to the nucleotide sequence of its partner strand.

## Structure of DNA Provides a Mechanism for Heredity

Genes carry biological information that must be copied accurately for transmission to the next generation each time a cell divides to form two daughter cells. Two central biological questions arise from these requirements: how can the information for specifying an organism be carried in chemical form, and how is it accurately copied? The discovery of the structure of the DNA double helix was a landmark in twentieth-century biology because it immediately suggested answers to both questions, thereby resolving at the molecular level the problem of heredity.

DNA encodes information through the order, or sequence, of the nucleotides along each strand. Each base—A, C, T, or G—can be considered as a letter in a four-letter alphabet that spells out biological messages in the chemical structure of the DNA., organisms differ from one another because their respective DNA molecules have different nucleotide sequences and, consequently, carry different biological messages. But how is the nucleotide alphabet used to make messages, and what do they spell out?

Figure: The nucleotide sequence of the human β-globin gene

It was known well before the structure of DNA was determined that genes contain the instructions for producing proteins. The DNA messages must therefore somehow encode proteins. This relationship immediately makes the problem easier to understand, because of the chemical character of proteins. The properties of a protein, which are responsible for its biological function, are determined by its three-dimensional structure, and its structure is determined in turn by the linear sequence of the amino acids of which it is composed. The linear sequence of nucleotides in a gene must therefore somehow spell out the linear sequence of amino acids in a protein. The exact correspondence between the four-letter nucleotide alphabet of DNA and the twenty-letter amino acid alphabet of proteins—the genetic code—is not obvious from the DNA structure, and it took over a decade after the discovery of the double helix before it was worked out.

The complete set of information in an organism's DNA is called its genome, and it carries the information for all the proteins the organism will ever synthesize. (The term genome is also used to describe the DNA that carries this information.) The amount of information contained in genomes is staggering: for example, a typical human cell contains 2 meters of DNA. Written out in the four-letter nucleotide alphabet, the nucleotide sequence of a very small human gene occupies a quarter of a page of text while the complete sequence of nucleotides in the human genome would fill more than a thousand books the size of this one. In addition to other critical information, it carries the instructions for about 30,000 distinct proteins.

This gene carries the information for the amino acid sequence of one of the two types of subunits of the hemoglobin molecule, which carries oxygen in the blood. A different gene, the α-globin gene, carries the information for the other type of hemoglobin subunit (a hemoglobin molecule has four subunits, two of each type). Only one of the two strands of the DNA double helix containing the β-globin gene is shown; the other strand has the exact complementary sequence. By convention, a nucleotide sequence is written from its 5′ end to its 3′ end, and it should be read from left to right in successive lines down the page as though it were normal English text. The DNA sequences highlighted in yellow show the three regions of the gene that specify the amino sequence for the β-globin protein.

At each cell division, the cell must copy its genome to pass it to both daughter cells. The discovery of the structure of DNA also revealed the principle that makes this copying possible: because each strand of DNA contains a sequence of nucleotides that is exactly complementary to the nucleotide sequence of its partner strand, each strand can act as a template, or mold, for the synthesis of a new complementary strand. In other words, if we designate the two DNA strands as S and S′, strand S can serve as a template for making a new strand S′, while strand S′ can serve as a template for making a new strand S. Thus, the genetic information in DNA can be accurately copied by the beautifully simple process in which strand S separates from strand S′, and each separated strand then serves as a template for the production of a new complementary partner strand that is identical to its former partner.

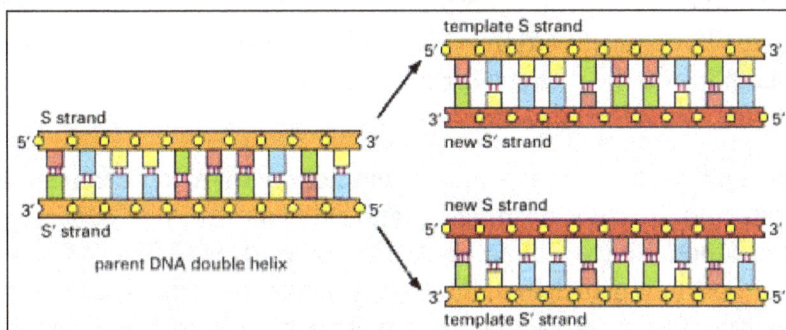

Figure: DNA as a template for its own duplication.

As the nucleotide A successfully pairs only with T, and G with C, each strand of DNA can specify the sequence of nucleotides in its complementary strand. In this way, double-helical DNA can be copied precisely.

The ability of each strand of a DNA molecule to act as a template for producing a complementary strand enables a cell to copy, or replicate, its genes before passing them on to its descendants.

## Eucaryotes: DNA is Enclosed in a Cell Nucleus

Nearly all the DNA in a eucaryotic cell is sequestered in a nucleus, which occupies about 10% of the total cell volume. This compartment is delimited by a nuclear envelope formed by two concentric lipid bilayer membranes that are punctured at intervals by large nuclear pores, which transport molecules between the nucleus and the cytosol. The nuclear envelope is directly connected to the extensive membranes of the endoplasmic reticulum. It is mechanically supported by two networks of intermediate filaments: one, called the nuclear lamina, forms a thin sheetlike meshwork inside the nucleus, just beneath the inner nuclear membrane; the other surrounds the outer nuclear membrane and is less regularly organized.

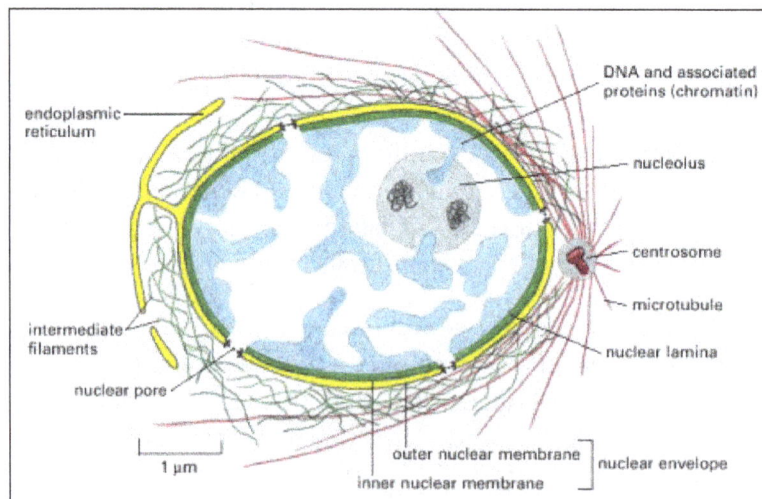

Figure: A cross-sectional view of a typical cell nucleus.

The nuclear envelope consists of two membranes, the outer one being continuous with the endoplasmic reticulum membrane. The space inside the endoplasmic reticulum (the ER lumen) is colored yellow; it is continuous with the space between the two nuclear membranes. The lipid bilayers of the inner and outer nuclear membranes are connected at each nuclear pore. Two networks of intermediate filaments (green) provide mechanical support for the nuclear envelope; the intermediate filaments inside the nucleus form a special supporting structure called the nuclear lamina.

The nuclear envelope allows the many proteins that act on DNA to be concentrated where they are needed in the cell, and, it also keeps nuclear and cytosolic enzymes separate, a feature that is crucial for the proper functioning of eukaryotic cells. Compartmentalization, of which the nucleus is an example, is an important principle of biology; it serves to establish an environment in which biochemical reactions are facilitated by the high concentration of both substrates and the enzymes that act on them.

# Genomic Organization

DNA is the genetic material containing all the information essential for life and the basis for heredity. The human genome is divided into 46 DNA molecules, or chromosomes, consisting of pairs of chromosomes 1 to 22 (autosomes), numbered sequentially according to their size, and of two sex chromosomes that determine whether an individual is male or female. Together, these molecules contain over 6 billion letters that when joined would measure ~2 m in length. It stands to reason that the human genome must be extensively packaged in order to fit inside the nucleus, the size of which is in the micrometre range.

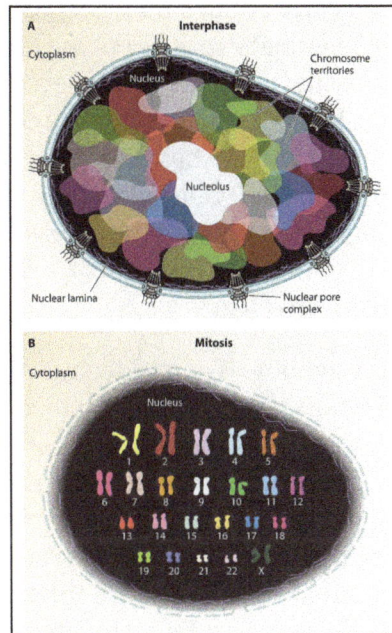

**FIG**

Human genome organization in a three-dimensional nucleus: (A) Chromosome territories observed during interphase. Nuclear pore complexes are shown perforating the nuclear envelope. The nucleolus is shown in white. The nuclear lamina is represented as a filamentous mesh inside the double nuclear membrane. (B) Example of a normal female karyotype as would be observed by SKY of mitotic cells.

The physiological state of genomic DNA is in the form of chromatin, where it is bound to histone and no histone proteins. Histones are by far the most abundant proteins in chromatin and bind DNA mainly as nucleosomes composed of two copies each of H2A, H2B, H3, and H4. Wrapping of DNA around nucleosomes represents the first level in packaging, which effectively shortens the length of chromosomes by 7-fold. Histones, particularly their amino and carboxy-terminal tails, are subject to posttranslational modifications (PTMs) on multiple residues, including methylation, acetylation, phosphorylation, sumoylation, ADP-ribosylation, or ubiquitinylation. PTMs regulate the activity of underlying genomic regions by altering how nucleosomes interact with each other and the DNA, thereby controlling access to given sequences, and by recruiting effector proteins that bind PTMs directly and interpret whether a region should be active or not. As such, chromatin could be considered the basic regulatory unit of genomes, and further packaging within the confines of the three-dimensional (3D) nuclear space can have a direct impact on its activity. The importance of three-dimensional chromatin organization both for reducing chromosome size and for other genome functions such as transcription is indeed recognized.

In healthy cells, higher-order chromatin organization is necessarily consistent with genome function and regulation. This level of organization is poorly described, with even the fundamental principles guiding interphase chromatin folding and unfolding still being unknown. How the genome folds is particularly important for transcription? Because control DNA elements and their target genes are not always next to each other along the linear genome sequence, a fact which has been apparent since Barbara McClintock's early studies on transposition. More recently, it was found that gene regulation by distal control elements such as enhancers is often associated with physical contacts between them. Given that tissue specificity is achieved through the combined action of regulatory sequences on target genes, this observation raises a compelling conundrum about specificity: if elements like enhancers can act long distance on genes located anywhere in the genome, how is specificity achieved? Three-dimensional genome organization appears to play an important role in this process by both promoting and restricting the access of control DNA elements to genes.

## Genome in Three-dimensional Nucleus

The Nuclear Lamina in mammals, the genome is contained within the cell nucleus, a double-membrane organelle that effectively segregates the transcription machinery from the cytoplasm, where protein production occurs. At its lowest resolution, genome organization is guided by contacts with several nuclear substructures. The nuclear envelope (NE) and its lamina are such structures. The outer nuclear membrane (ONM) and inner nuclear membrane (INM) of the NE are populated by nuclear envelope trans membrane proteins (NETs), which associate with the lamin proteins on the INM face to form the nuclear lamina. The lamins are intermediate filament proteins, which, together with NETs, can bind many different proteins, including chromatin components such as heterochromatin protein 1 (HP1) and histones. Interactions between the lamina and the chromatin can regulate the position of chromosomes in the nucleus and various other activities.

Chromatin interacts with the nuclear lamina through lamin-associated domains (LADs) that vary in size from 0.1 to 10 Mb and frequently contain transcriptionally inactive heterochromatin. LADs are generally conserved but can also be cell type specific. While conserved LADs span regions with very low GC content and are gene poor, cell type-specific LADs usually have a higher GC content and correlate with tissue-specific gene expression. These observations suggest that inactive chromatin regions, even those that are gene rich, tend to localize at the nuclear periphery or the nucleolus. In fact, simply relocating a given region to the nuclear lamina is often sufficient to reduce gene expression, but this is not always the case.

## Nuclear Pore Complex

The nuclear pore complex (NPC) is another nuclear substructure involved in regulating gene expression that might play a role in chromatin organization. NPCs are evolutionarily conserved structures that mediate all transport between the nucleus and the cytoplasm. They are very large, ranging in mass from ~60 million to 100 million Da, depending on the organism, and exhibit an 8-fold symmetrical/cylindrical geometry around a central transport channel. NPCs "perforate" the two lipid bilayers of the nuclear envelope and are composed of >30 different nucleoporin proteins. Nucleoporins either are part of the core (integral proteins), form filaments extending from the NPC core toward the cytoplasm or nucleoplasm (those containing FG repeats, or FG-Nups), or are part of the nuclear fibers (nuclear basket).

The chromatin environment around the nucleoplasmic face of NPCs differs from that of the rest of the inner nuclear membrane despite their physical proximity to the lamina. Early electron microscopy studies of nuclei in higher eukaryotes revealed the presence of heterochromatin exclusion zones (HEZs) around NPCs. These gaps in the heterochromatin landscape of the inner membrane vary in size, tend to be cone-like, and are populated by euchromatin that extends from the NPC fibers to the nucleoplasm. Interestingly, contacts between nucleoporins and chromatin could be captured by double cross-linking. Although direct physical NPC-chromatin interactions cannot be concluded from these experiments, they nonetheless suggest an intimate link between chromatin and the nuclear transport machinery. Accordingly, several nucleoporins have been involved in the activation of transcription, and although a large fraction of nucleoporins is found free in the nucleoplasm, NUP98 has been shown to bind genes on its own or as part of the NPC. Nucleoporins might also contribute to the compartmentalization of chromatin marks along human chromosomes given that they have been linked to insulation in yeast and are required for HEZ establishment in human.

## Nucleolus

A third type of nuclear landmark involved in genome organization is the nucleolus. Nucleoli are dense structures, visible by light microscopy, where rRNA synthesis and preribosome assembly occur. They form around grouped rRNA genes from different chromosomes that are transcribed by RNA polymerase I (Pol I) and are located where nascent rRNA transcripts are processed and packaged into preribosomes. Several RNA polymerase II genes were found to copurify with the nucleolus when transcriptionally inactive. These nucleolus-associated domains (NADs) significantly overlap LADs and have similar GC-poor and gene-poor contents. Loci were actually found to colocalize with either NADs or LADs, suggesting that a certain amount of redistribution occurs between the two regions after mitosis and possibly that similar factors target the inactive chromatin to either the lamina or nucleolus. Other substructures have also been linked to chromosome organization and include Cajal and promyelocytic leukemia (PML) bodies.

### Visual and Molecular Analysis of Genome Organization

In addition to anchors with nuclear landmarks, human genome organization is guided by chromatin interactions within (cis) and possibly between (trans) chromosomes. These interactions are driven by the chromatin landscape and are thus often tissue specific and regulated. Our current view of genome organization is based largely on, and perhaps is limited by, data derived mainly from only two types of approaches. A variety of microscopy techniques, including several fluorescence in situ hybridization (FISH) procedures, visualized by conventional or super resolution light microscopy, is currently used to directly measure the proximity between DNA segments. These methods yield information-rich data about genome topography in individual cells that are well complemented by insights obtained from cell populations using molecular techniques such as chromosome conformation capture (3C) and its derivatives. This second type of technique infers

DNA proximity by quantifying the frequencies of contacts between DNA segments and considering them to be inversely proportional to their original distance in vivo. Representative methods from each category are presented below.

## Visualizing Genome Organization

Until the advent of molecular techniques such as 3C and its high-throughput derivatives, the predominant method for determining nuclear organization and chromatin conformation was FISH. This cytogenetic approach has been used for a variety of applications, from clinical diagnostics to the study of genome architecture. Sensitivity and resolution are limiting factors to consider when designing a FISH experiment. Sensitivity depends on the light-capturing capability of a particular microscope, therefore determining the size of the probe (larger probes will generally produce stronger signals). Probe size then brings in one aspect of resolution: being able to distinguish between two points along the length of a chromosome. For example, fosmid probes are frequently used to measure chromatin compaction or to identify the colocalization of genes with remote regulatory elements. However, due to their size (~40 kb), genomic distances of <100 kb cannot be resolved. However, oligonucleotide-based probes of ~10 kb in size with high hybridization efficiencies that produce strong signals allow the chromatin conformation of sub-100-kb genomic regions to be determined. There is also the resolution of the light microscope to consider conventional light microscopy cannot resolve structures with sizes of <200 nm in the x and y planes and <500 nm in the z plane. However, superresolution microscopy is bringing the light diffraction limit down to the tens-of-nanometers scale.

## 2D-, 3D- and Cryo-FISH

The observation that complementary nucleotide sequences could hybridize to each other and form more stable complexes than noncomplementary sequences was the basis for the first in situ hybridization analysis that identified the position of ribosomal DNA within the nucleus of a frog egg. Molecular cytogenetics is based on this procedure, with the replacement of radioactive labels with more stable fluorochromes providing improved safety and ease of detection. The FISH technique, then, relies upon probe sequences that target genomic DNA, which either are directly labeled with a fluorochrome or have been modified to contain a hapten (such as biotin) and are then rendered fluorescent indirectly by enzymatic or immunological detection. Target and probe DNAs must go through a denaturation step to allow target-probe hybridization. The fluorescently labeled genomic loci can then be visualized by using a fluorescence microscope.

There are now a diverse number of FISH assays that can be applied to megabase (metaphase chromosomes), submegabase (interphase chromosomes), and even nucleotide (oligonucleotide arrays) resolutions. The two-dimensional FISH (2D-FISH), 3D-FISH, and cryo-FISH variations of the FISH process have been used to directly visualize and measure the nuclear distance between DNA segments, the nuclear location of DNA segments or indeed whole chromosomes, and the location of a DNA segment in relation to the rest of the chromosome (i.e., within or "looped out" of the chromosome territory [CT]). In the 2D-FISH procedure, cells are fixed in methanol ascetic acid (MAA), which generates looser chromatin packaging due to the flattening out of the cells on the slide; however, results are comparable to those for paraformaldehyde (pFA) fixation of the same cells. The advantages conferred by 2D- over 3D-FISH are clearer visualization and rapid image analysis. This technique has been predominantly used for determining the nuclear location of genes or translocation from the nuclear periphery to the center and vice versa and for determining the location of a DNA segment in relation to the rest of the chromosome. We have used 2D-FISH to determine changes in chromatin condensation at the submegabase level, during differentiation and across a polarizing axis during development or between wild-type and mutant cells.

Not all direct analyses of chromatin structure can be adequately visualized by 2D-FISH. Colocalization of discrete genomic loci, for example, such as promoter-enhancer interactions, requires 3D reconstruction of nuclei. In 3D-FISH, cells or tissue sections are fixed in 2 to 4% pFA, and image capture requires confocal microscopy or deconvolution software if images are taken with a wide-field fluorescence microscope that has the capacity to generate image stacks through the z dimension. Applying this method to tissue sections and cell lines derived from embryonic day 10.5 (E10.5) limb buds of mouse embryos, we found that the colocalization frequency of Hoxd13, crucial for distal limb development, with a limb-specific long-range enhancer is increased in expressing cells. This type of analysis, in combination with 3C, has also been done on the Shh locus at the same stage of limb development and has been used for visualizing the colocalization of a single olfactory receptor allele and an enhancer element in individual sensory neurons. More recently, the Lomvardas laboratory identified the colocalization of multiple putative enhancers with individual olfactory receptor (OR) alleles by chromosome conformation capture-on-chip (4C) and 3D-FISH and determined by Hi-C that many of these colocalizations occurred in trans, which was also confirmed by FISH. Each olfactory sensory neuron expresses only one of 2,800 olfactory receptor alleles, and by generating a DNA FISH probe that simultaneously detected most OR loci, this group showed that the silent OR alleles converge to form exclusive heterochromatic foci in a cell type-specific and differentiation-dependent manner.

FISH has been combined with live-cell imaging to show that targeted transcriptional activation of a chromosome locus can induce movement from predominantly peripheral to more interior nuclear locations and that chromatin movement is restrained by the nuclear architecture. Live-cell imaging involves the incorporation of LacO or TetO arrays into genomic regions of interest and subsequent illumination through the binding of a vector cassette containing LacR/TetR with a fused fluorescent protein. Tagged loci can then be visualized, and their movement can be monitored while cells are maintained under suitable conditions. Live-cell imaging of a whole chromosome has also been achieved by combining LacO/LacI tagging with photoactivatable histones. Three-dimensional FISH has been used in conjunction with live-cell imaging and mathematical models to probe chromatin topography at the immunoglobulin heavy-chain locus. This combined approach has elucidated the compartmentalization and large chromatin changes that occur during B lymphocyte development and has generated a strong model for how the widely spread Igh coding elements (within three domains over >2 Mb) can frequently interact. Spatial confinement of the interacting domain was posited to be the main driver in genomic interactions between the Igh coding elements, which is a scenario that we identified by 3D-FISH combined with chromosome conformation capture carbon copy (5C) for increased gene enhancer colocalization.

The cryo-FISH technique has been applied for determining the spatial intermingling of interphase chromosome territories and for validating results from 4C technology that identified functionally significant long-range chromosomal interactions. Cryo-FISH involves pFA fixation and then embedding of cell pellets in sucrose before cells are frozen in liquid nitrogen. Ultrathin cryosections

(150 to 200 nm) can be generated to allow 2D, wide-field microscopy analysis of sequential sections through nuclei with no reduction in *z* resolution while improving the hybridization efficiency and preserving the chromatin ultrastructure.

## Inferring Genome Organization

Whereas distances between genetic loci can be measured directly in single cells by microscopy, the physical proximity of chromatin can also be deduced based on the frequency at which DNA segments interact with each other in cell populations in vivo.

This approach is based on the premise that interactions between close regions are more likely to be captured by cross-linking than are those between regions located far away and that the contact frequency over the cell population at a given time essentially reflects how chromatin is organized in the nucleus of individual cells. Genome architecture can be modeled with this type of data by considering the frequency to be inversely proportional to the physical distance. Several molecular techniques are available to quantify chromatin contacts, including 3C and 3C-related methods (3C technologies) and chromatin interaction analysis by paired-end tag sequencing (ChIA-PET). In contrast to FISH, where analysis is performed at the single-cell level, these techniques always capture contacts in cell populations and yield average structure models, with the exception of one study where single cells were analyzed.

## 3C

The "chromosome conformation capture" (3C) technique was developed by Dekker et al. >10 years ago and is routinely used to study the organization of short genomic regions at high resolution compared to the resolution of most visual techniques. During 3C, a population of cells is first chemically fixed with formaldehyde to create covalent bonds between chromatin segments. The cross-linked chromatin is then digested with a restriction enzyme, which cuts at specific sites across the genome. The type of enzyme selected defines the resolution of the 3C experiment, with those recognizing palindromes of 4 bp yielding higher-resolution libraries (256 bp) than 6-cutters (i.e., enzymes that recognize 6-bp sequences; 4,096 bp). The digested DNA is next diluted, and ligase is added to join cross-linked fragments pairwise. This step generates unique DNA junctions measurable by various PCR methods.

The original 3C method is outlined from top to bottom on the left. Formaldehyde cross-linking captures interactions between DNA segments (blue and green lines) mediated by protein complexes (colored shapes). The chromatin is next digested with a restriction enzyme, and the free DNA ends are joined by proximity ligation before reverse cross-linking and purification. The genome-wide ChIA-PET and Hi-C techniques are related to 3C, and key steps are shown from left to right. The Y-shaped molecule represents antibodies. Biotinylated nucleotides are shown as red dots. Streptavidin beads are shown in brown. The genome-scale 4C and 5C methods indicated at the bottom require the production of 3C libraries, and specific key steps are outlined from left to right. Green arrows represent PCR primers specific to the bait region. 5C primers used during the ligation-mediated amplification step are illustrated with green and blue lines, where the light and dark gray moieties represent universal primer sequences.

As interactions are measured individually, 3C is generally used for small-scale analysis and was first applied to confirm the Rabl-like organization of chromosomes in *Saccharomyces cerevisiae*. It was then used to explain long-range regulation at the β-globin cluster during erythroid differentiation in studies that followed the very first demonstration of enhancer-promoter looping at the rat prolactin gene. Long-distance cis and trans physical contacts have since been found genome-wide and regulate the activity of enhancers at promoters, insulator function, transcriptional silencing, imprinting, and X inactivation.

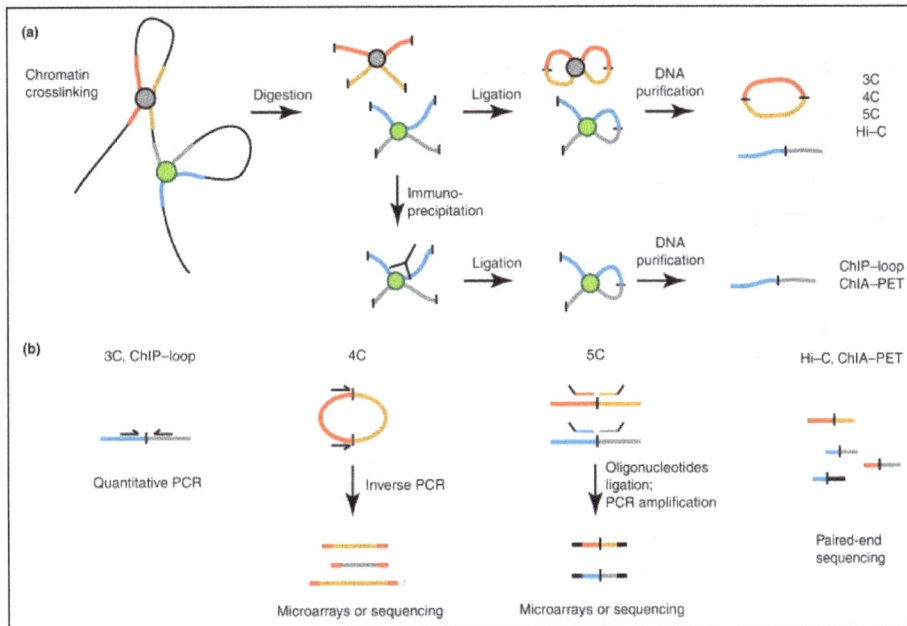

Inferring chromatin organization.

Although the data generated with 3C and its related technologies have largely been corroborated by other methods like FISH, it appears that we still have much to learn about how these methods work. For instance, it was originally assumed that dilution of the 3C reaction mixture prior to ligation was required to favor the ligation of cross-linked restriction fragments. Recent work shows that this is likely not the case, since a substantial portion of the digested DNA remains trapped within the cross-linked nuclei. The nucleus-bound DNA was found to actually contribute most of the measured 3C signal, indicating that ligation occurs mainly while the DNA is still bound to the nuclei rather than in solution. Dilution of 3C reaction mixtures might therefore simply be required to decrease SDS concentrations prior to ligation. In another study, it was found that enhancer-promoter ligation products actually represent <1% of all the restriction fragments subjected to ligation. This result likely reflects the frequency of interactions between regulatory DNA elements in vivo; the many different types of products generated at the ligation step; as well as the efficiency, specificity, and stability of the formaldehyde cross-links formed during a 3C experiment. Also, we recently demonstrated that data from FISH and 5C are sometimes discordant at high resolution, suggesting that parameters other than distance might influence the interaction frequency.

## 4C

The chromosome conformation capture-on-chip (4C) techniques were the first set of methods designed to improve the throughput and resolution of 3C. Each technique was developed independently to identify all contacts between a given region and the rest of the genome without any prior knowledge of the contacting domains. Generally, these methods work through the generation of very short ligation products between a specific restriction fragment (the "bait" or "anchor") and the rest of the genome, which are quantified by using either microarrays or sequencing.

The "open-ended" 3C methodology was used to identify HoxB1-associated loci throughout the genome in mouse embryonic stem (ES) cells. This approach identified interacting DNA segments by sequencing of short ligation products amplified from the 3C library by inverse PCR with nested HoxB1 primers. The Hoffman group developed the "associated chromosome traps" (ACT) technique to find genomic domains that interact with the mouse insulin-like growth factor (Igf2)/H19 imprinting control region. This method differed from the open-ended 3C technique in that short 3C ligation products were first ligated to linkers before amplification with linker primers and nested Igf2/H19-specific primers. The "circular chromosome conformation capture" approach used the mouse H19 imprinting control region (ICR) as a general model to introduce the technique. It was also similar to the open-ended 3C technique but used custom arrays specific to the 4C libraries to quantify novel chromatin interactions. In contrast, the 4C method used microarrays containing genome-wide probes, which were later phased out in favor of the more sensitive high-throughput sequencing.

The chromosome conformation capture-on-chip method was the only easily scalable 4C approach. Here, ligation products from a 3C library are further digested with a restriction enzyme that cuts more frequently and are religated into circular DNA. PCR primers designed to face outwards on either side of the bait are then used to simultaneously amplify all interacting fragments, and the amp icons are quantified by deep sequencing. The application of sequencing greatly increased the scale and sensitivity of the 4C assay, allowing genome-wide profiling of interactions with the bait region. No other 3C-related technique can yet generate comparable high-resolution interaction profiles. 4C was also modified to include an immunoprecipitation step before the first ligation reaction, which enables the capture of chromatin interactions mediated by specific proteins.

## 5C

Since 4C elucidates only interactions between a single restriction fragment and the rest of the genome, it cannot be used to predict the conformation of entire domains or chromosomes. The chromosome conformation capture carbon copy (5C) method, on the other hand, is suitable for this type of analysis, as it can detect up to millions of 3C ligation junctions between many restriction fragment pairs simultaneously.

5C was designed to increase the throughput and accuracy of 3C by combining 3C with a modified version of the ligation-mediated amplification (LMA) technique. For 5C, a series of primers is computationally designed at the restriction site of each fragment in the region of interest. These 5C primers are then pooled with a 3C library where they anneal to targeted 3C fragment ends. Primers located next to each other across the 3C junction are next ligated together by Taq ligase, generating new synthetic DNA molecules. This 5C library is amplified by using the common tails present in the 5C primers and is quantified by using high-throughput sequencing. This process results in the interrogation of all chromatin interactions between fragments represented by 5C primers. As the LMA step quantitatively ligates primers onto 3C junctions, the 5C library is essentially a "carbon copy" of existing 3C products.

5C can be used at different scales to probe various biological questions. That it uses predefined primer sets to measure chromatin contacts, however, implies that 5C can measure contacts only for regions covered by the primer library. 5C has been used to study chromosome-sized regions at high resolution by multiplexing large numbers of primers. We determined the physical organization of the human Hox clusters in various cell systems with 5C and used the data to model spatial chromatin organization computationally. Other groups have used 5C to study the alpha-globin cluster, the three-dimensional organization of the bacterial Caulobacter crescentus genome, the regulatory landscape of mouse X inactivation, and changes in developmentally regulated chromatin domains.

## Hi-C, GCC and TCC

The Hi-C technology, sometimes called genome-wide chromosome conformation capture, uses high-throughput sequencing to directly quantify proximity ligation products in contact libraries and therefore can be used to probe the spatial organization of an entire genome. The type of data produced with Hi-C is therefore more comprehensive than what other 3C-type methods usually yield. The production of Hi-C libraries is similar to the production of 3C libraries: cell populations are first chemically fixed with formaldehyde, and the chromatin is digested with a restriction enzyme or DNase, which was recently reported to achieve higher resolutions. The restriction fragment overhangs are then filled with Klenow enzyme and a mixture of deoxynucleoside triphosphates (dNTPs) that includes biotin-14-dCTP. This is followed by blunt-end ligation with T4 DNA ligase to join cross-linked DNA fragments, reverse cross-linking, and purification. These unprocessed Hi-C libraries are then sheared by sonication and size selected prior to pulldown on streptavidin-coated beads, which enriches the samples for DNA sequences containing the informative ligation junctions.

Illumina's paired-end sequencing has so far been the method of choice to identify the sequences on either side of Hi-C junctions, but longer read lengths will increasingly make single-readthrough approaches viable alternatives for Hi-C analysis. Current methods map each side of the sequence reads separately to a reference genome, where the quantities of all valid Hi-C pairs are catalogued in a matrix format spanning the entire genome. Since the complexity of Hi-C libraries is very high, a sizeable amount of sequencing is required. The data are typically binned at different sizes based on sequence depth and library quality. The organizations of large genomes such as those of human and mouse have consequently been described mainly at resolutions ranging from 40 kb to 1 Mb, except for a recent study where the Hi-C protocol was modified to achieve kilobase resolutions. Hi-C data have been used to correlate genome architecture with several genomic features such as replication timing. It was found that differential firing at origins could be explained by the spatial compartmentalization of origins into different units of three-dimensional chromatin architectures. Hi-C was also used to show that the proximity of chromatin correlates well with the incidence of intra- and interchromosomal translocations from double-stranded DNA breaks. The generation of very-high-resolution Hi-C data sets is prohibitively expensive at present, and until sequencing costs decrease or alternative Hi-C protocols are developed, approaches such as 4C and 5C remain ideally suited for high-resolution analyses.

Two techniques similar to Hi-C were reported at around the same time as the original Hi-C report and were used to explore the genome organization of the yeast Saccharomyces cerevisiae. The "genome conformation capture" (GCC) technique reported a global map of chromosomal interactions in yeast by shearing and directly sequencing conventional 3C libraries. Since the majority of the DNA in 3C libraries does not contain any 3C ligation junctions, this approach yielded a small number of usable reads that was nonetheless sufficient to identify contact networks in the small S. cerevisiae genome. GCC has since been used to map the spatial organization of the Escherichia coli nucleoid. Another group developed a 4C-inspired approach, which involved the ligation of biotinylated adaptors to mark ligation junctions. This study generated kilobase-resolution maps that were used to construct a three-dimensional model of the yeast genome.

Hi-C itself has been modified into the "tethered conformation capture" (TCC) method, which generates the same chromatin interaction data but boasts higher signal-to-noise ratios. The premise of TCC is that random intermolecular ligation products between non-cross-linked DNA fragments are the largest source of noise and appear most frequently as interchromosomal contacts in the data. In TCC, the frequency of this event was decreased by biotinylating the cross-linked protein-DNA complexes and tethering them to streptavidin-coated magnetic beads prior to the Hi-C ligation step. This modification was suggested to improve the efficiency of ligation between cross-linked DNA strands compared to that of traditional Hi-C ligation under diluted conditions. Given that most contacts were recently shown to form in the nucleus-bound fraction instead of in solution, it will be interesting to see how immobilization of complexes on beads decreases the incidence of nonspecific ligation products. Comparable signal-to-noise ratios have since been obtained with the original Hi-C procedure, but the fact that the levels of random ligation products tend to vary significantly between experiments nonetheless highlights the importance of optimizing this aspect of library production to improve read depth.

## ChIA-PET

While Hi-C can be used to identify contacts genome-wide, it does not provide information about the nature or the function of these interactions. The "chromatin interaction analysis by paired-end tag sequencing" (ChIA-PET) method was developed to probe this type of question by mapping chromatin networks associated with specific proteins. ChIA-PET is a genome-wide technique that uses a chromatin immunoprecipitation (ChIP) step to isolate interactions between all regions bound by a particular protein. Like Hi-C, ChIA-PET identifies contacts in cell populations fixed with formaldehyde, but the fixed cells are sonicated and used first for ChIP of the protein of interest. Biotinylated DNA linkers are next added at the ends of the coimmunoprecipitated DNA segments, and the resulting cross-linked DNA fragments are ligated together intramolecularly. The ChIA-PET junctions generated by this process are then excised with restriction sites featured in the linkers prior to purification on streptavidin beads and paired-end sequencing. Each half of the ChIA-PET products is finally mapped to a reference genome and joined to reveal the location of protein-mediated chromatin contacts. This method has been used to map networks associated with RNA polymerase II, CCCTC-binding factor (CTCF), and the estrogen receptor genome-wide.

## Chromosome Organization in the Nuclear Space

A general model of genome architecture wherein chromosomes are organized in hierarchical length scales has recently emerged. From low to high resolution, chromosomes first fold to occupy distinct territories and positions in the nuclear space defined in part by interactions with nuclear subdomains, including heterochromatic regions. Individual chromosomes are then folded into compartments A (open/active) and B (closed/silent) that preferentially interact together, respectively. Within compartments, the chromatin is packaged in the form of topologically associated domains (TADs), largely conserved between cell types and across species. The chromatin is further folded into sub-TADs, the topologies of which can vary in a tissue-specific manner. Ultimately, genomic DNA is wrapped around nucleosomes, which represents the first level of genome folding; however, how DNA is packaged between the resolution of the 10-nm fiber and sub-TAD scales is still largely unknown. Below, we explore the organization of chromatin across genomic scales, from chromosome territories to individual genes.

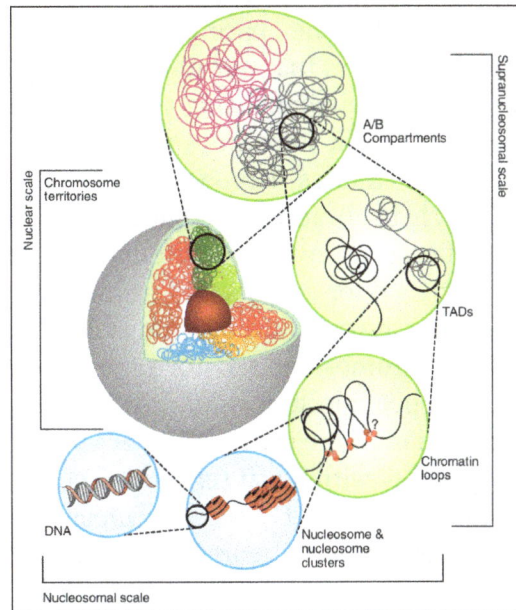

Chromatin organization across genomic scales.

The chromatin fiber from one chromosome is unraveled to illustrate four different organization levels. Chromatin conformations are presented from low (top) to high (bottom) resolutions. The chromatin fiber and corresponding chromosome territory are shown in pink. A and B compartments (multimegabase scale) are shown separately to highlight their inherently distinct nature, although there is no evidence that their conformations differ at the level of TADs (megabase scale). Three examples of chromatin looping (submegabase scale) are shown: (i) enhancer-promoter, (ii) enhancer-silencer, and (iii) insulator-insulator. E, enhancer; P, promoter; S, silencer; I, insulator.

## Chromosome Territories

The work of Carl Rabl and Theodor Boveri suggested long ago that animal interphase chromosomes adopt a form of territorial organization where interchromosomal contacts are minimized. Chromosome territories (CTs) were visualized under a light microscope in these early studies and identified while the movement of DNA during the cell's life cycle was tracked. More specialized approaches using UV irradiation and pulse labeling later supported the existence of CTs by demonstrating the preferential distribution of DNA aberrations within chromosomes. FISH has since been used to visualize the location of chromosomes and clearly demonstrates their propensity to form individual domains. 3C-based data corroborate the existence of CTs. For instance, 4C analyses and genome-wide Hi-C studies capture more intrachromosomal contacts than interactions between chromosomes, even for loci hundreds of megabases apart on a given chromosome. Accordingly, Heride et al. demonstrated by 3D-FISH that homologous chromosomes in a diploid cell are far apart from each other. Recent work on haplotype reconstruction using Hi-C data supports these findings by demonstrating that chromosome haplotypes in diploid cells do not interact frequently with each other.

Although chromosomes mostly keep to themselves, they can considerably interact with other CTs. For instance, contacts between small, gene-rich chromosomes in Hi-C libraries of human lymphocytes were shown to occur more frequently than would be expected based on their size. Several loci were shown to loop out of their chromosome territory, coinciding with both an open conformation and active expression and suggesting that the space between chromosome territories might be important. It is important to note, however, that our understanding of the structure and biology of CTs is derived largely from FISH experiments using probe sets that do not cover entire chromosomes, and thus, such looping out might sometimes reflect only extrusion from the visualized regions rather than the actual CT. Nevertheless, comparison of the FISH signals from conventional whole-chromosome "painting" (i.e., hybridization with fluorescently labeled chromosome-specific probes) to those from exome painting of the entire chromosome revealed that chromatin segments at the surface of CTs are enriched for exons, residing largely away from the more compact CT core, which is consistent with looping out. Several groups demonstrated a significant amount of intermingling between different chromosome territories on a cell-specific basis, although the extent of these contacts remains an open question. Similarly, interchromosomal contacts between a select set of highly transcribed regions were captured by TCC, and it was suggested that access to the transcription machinery, possibly within transcription factories, can drive the formation of contacts. Other studies using genome-wide Hi-C data from mouse and human show that physical proximity prior to chromosomal rearrangement correlates well with the incidence of translocations genome-wide.

Chromosomes have preferred radial positions in the nucleus of mammalian cells. 3D-FISH and chromosome painting analyses in various cell types showed that chromosomes tend to localize at either the nuclear center or the periphery according to gene density. Human chromosome 19, for example, is a small gene-rich chromosome more frequently found at the center of the nucleus than chromosome 18, which is similar in size but gene poor. This behavior has been observed for multiple species, including other primates, rodents, cattle, and birds, and thus appears to reflect a general feature of eukaryotic nuclear organization. Accordingly, Hi-C analysis showed that all small gene-rich human chromosomes interact more frequently with each other than with the similarly sized chromosome 18. Computational modeling of TCC data further indicated that gene-dense chromosomes tend to localize to the center of the nucleus, while a group of gene-poor chromosomes localized to the periphery. Other studies argue that transcription activity, chromatin remodeling, replication timing, chromosome size, GC content, and gene density within megabase-sized genomic windows contribute to the preferential position of chromosomes relative to the nucleus center.

Whether the position of a specific chromosome and its interacting partners is functionally important is unknown. The position of CTs was shown to be cell type specific, suggesting that boundaries shared between a given chromosome and its neighbors, along with their relative nuclear position, might be functionally relevant. However, while the CT position tends to be stable within a given cell, it can vary significantly from cell to cell, indicating that not all trans contacts are required and that an exact nuclear position may not be essential. This is supported by the observation that tethering to the nuclear periphery mostly alters the expression of genes in cis, with little effect on transcription from other chromosomes, even though trans contacts must be altered in these experiments. The positions of CTs might be influenced by their internal chromatin organization. Chromosome folding into a CT was suggested to act as a barrier to internal nuclear movement, and accordingly, genes embedded deep within a chromosome territory are more difficult to activate.

While the function of CTs as a first level of compartmentalization is recognized, there is at least one example where they can serve a very different role. It was found that the CT position in the rod photoreceptor cells of nocturnal mammals is inverted relative to the conventional architecture seen in diurnal animals and most eukaryotic cells. In nocturnal retina rod cells, the heterochromatin localizes at the center of the nucleus, and the euchromatin lines the nuclear periphery. Computational modeling of this nuclear organization in the eye indicates that rod nuclei can act as collecting lenses that efficiently channel light in this configuration, thus contributing to an adaptation to the nocturnal life-style.

## Chromosome Compartments

The original Hi-C study reported the genome-wide chromatin organizations of two human cell lines at a resolution of 1 Mb. The resulting Hi-C contact matrices displayed a type of checkerboard-like contact pattern where multimegabase regions interact even across large distances along chromosomes. Principal component analysis (PCA) of the Hi-C data was used to segregate contact frequencies into pairwise states, which uncovered the existence of two types of chromosome compartments. The open "A" compartments include regions with high GC content that are enriched in genes, transcription activity, DNase hypersensitivity, and histone modifications associated with active chromatin (H3K36me3) and poised chromatin (H3K27me3). In contrast, B compartments show higher interaction frequencies, a stronger tendency toward self-association, and high levels of the silencing H3K9me3 mark. The position of B compartments was also found to be highly correlated with late replication timing and LADs, suggesting a proximity to the nuclear periphery not observed for A compartments.

The segregation of CTs into A and B compartments has thus far been observed for all autosomes and for all mammalian cell types examined. Their distribution along chromosomes is also highly stable across cell types but can vary, pointing to a regulatory role. However, contacts within compartments tend to be weak and spread over large groups of restriction fragments throughout the domains, suggesting that compartments may exist only transiently or may even form simply as a consequence of shared features.

## TADs and Sub-TADs

While exploring chromosome organization at smaller scales using 5C and Hi-C, blocks of dense chromatin were identified in human, mouse, and *Drosophila melanogaster*, which interact more frequently within themselves than with neighboring regions. The existence of topologically associating domains (TADs) is supported by the finding that FISH probes intermingle more frequently within TADs than between them. TADs are clearly visible in 5C and Hi-C data and are defined by sharp changes in the contact frequency from one region to the next. TADs are thought to partition genomes into distinct globular units that can remain spatially distant even if they are adjacent along chromosomes. Perhaps because of their small genome sizes, TADs have not been identified in bacteria or yeast. They are also not found in the larger plant genomes, where other types of chromatin domains exist.

The first identified TADs had an average size of between 0.5 and 1 Mb and were argued to have distributions that are highly conserved between cell types and species. Similar-sized domains had previously been observed by microscopy; The domains were visible at the edges between pairs of chromosome territories and were suggested to represent basic CT building blocks. The fact that both TADs and CT domains exist at approximately the same scale suggests that they may be the same chromatin structures.

The actual molecular makeup of TADs and the mechanisms by which they are formed are the subjects of much investigation. Genes within TADs tend to be co-expressed during differentiation, and the position of TADs correlates well with the distribution of activating and repressing histone marks. The boundaries delineating the original TADs were found to be enriched in transcription start sites (TSSs), active transcription and the corresponding histone marks, housekeeping genes, tRNA genes, and short interspersed nuclear elements (SINEs). TAD boundaries are also enriched in binding sites for the architectural proteins CTCF and cohesin. The fact that deletion of the sequence at a TAD boundary results in a partial fusion of the domains and affects nearby gene expression highlights the importance of this level of compartmentalization in the regulation of genes. Also, the observation that the positions of TADs viewed at the megabase scale are largely conserved suggests that sequence-related factors either within TADs or at their boundaries are involved in their formation. This is supported by the observation that most long-range gene regulation by enhancers is constrained within TADs.

High-resolution 5C analysis later demonstrated that TADs are further divided into submegabase-sized structures that are loosely referred to as "sub-TADs". Finer substructures were actually already visible within the original lower-resolution TAD data but were subsumed owing to lower confidence. The average size and position of TADs appear to depend on both data resolution and the parameters used to identify them. TADs are usually defined computationally by the position of their boundaries, identified with a directionality index (DI). The DI is a measure of the difference between upstream and downstream interactions along a chromosome, and boundaries are defined as locations where high contact frequencies shift from downstream to upstream regions. As such, DI data can vary significantly depending on the sliding window size selected. Whereas small window sizes return smaller TADs, larger ones yield larger TADs that often contain groups of smaller domains. In fact, when the original Hi-C data from which megabase-sized TADs were first identified were reanalyzed with a different algorithm that uses smaller window sizes, it was found that the larger conserved TADs tend to consist entirely of smaller domains with an average size of 0.2 Mb. These domains were stable between cell lines and persistent across resolutions, and their boundaries were also enriched in CTCF binding and activating histone marks.

The finding that TADs can be substratified into smaller domains displaying a pronounced hierarchical organization suggests that topologies from different length scales interact with each other to yield functional genome architectures. However, how domains at the sub-TAD scale relate to each other and the more conserved TADs is unknown. The notion that domains might sequentially interact with each other in progressively larger structures challenges the functional significance of domain classification based on size. In contrast to TADs, chromatin organization at the submegabase scale was found to be more tissue specific and mediated by various protein complexes. Whereas invariant subdomains relied on CTCF and the cohesin complex forming long-range interactions, the more tissue-specific enhancer/promoter contacts within and across subdomains required mediator and cohesin. The unique topological signatures of sub-TADs therefore appear to reflect the levels and types of genomic activities. Whether boundaries at the invariant subdomains are functionally distinct from the ones characterized at larger TADs is unclear, and it will be interesting to see if the structures defined by them are one and the same.

## Chromatin Looping and Looping Out

The finer structures observed at the submegabase scale in high-resolution conformation data highlight the existence of long-range contacts that either form the base of stable domains or are directly involved in regulating processes such as transcription. These long-distance interactions reflect the ability of chromatin fibers to fold into "loops." Chromatin looping was evident as early as 1878, when Walther Flemming first reported the existence of "strange and delicate structures" in the nucleus of amphibian oocytes. It was J. Ruckert, however, who concluded that they were looped chromosomes, calling them "lampbrush chromosomes" for their resemblance to the bristled brushes then used to clean the soot off oil-burning lamps. A detailed visual and biochemical account of chromatin loops came only some 50 years later through the work of Joseph Gall and facilitated the discovery of DNA loops in human cells, most of which are several orders of magnitude smaller than the very large loops of the lampbrush chromosomes containing upwards of several hundred kilobases of DNA.

As it relates to transcription in mammals, the function of chromatin looping appears chiefly to either promote or prevent contacts between gene promoters and regulatory elements, particularly enhancers. Since their initial discovery in viral genomes, cellular enhancers have been found genome-wide and are largely responsible for the tissue-specific expression of genes. Despite the fact that enhancers and promoters can each initiate bidirectional transcription, enhancers can activate transcription in either orientation, whereas promoters cannot. Enhancers can also drive transcription from a position upstream of, downstream of, or within target genes. Enhancers are found at various distances from the promoters that they regulate and can work over very long distances in *cis* or even from different chromosomes. The observation that chromosome territories intermingle significantly in the nuclei of human cells supports the existence of *trans*-regulatory mechanisms, although the functional relevance of most contacts with respect to transcription remains largely undefined.

A striking example of long-range transcription regulation was described for the mouse Sonic hedgehog (Shh) gene, which is activated during limb development by an enhancer known as the zone of polarizing regulatory sequence (ZRS; also known as MFCS1), imbedded within the intron of another gene located >1 Mb away. 3C analysis revealed that Shh activation by the ZRS correlates with physical contacts between them, suggesting a looping of the chromatin path between the enhancer and Shh. A study that combined genetic manipulation of the enhancer locus, transgenics, and 3D-FISH showed that while the 5′ end of the ~800-bp ZRS is sufficient to drive an adjacent reporter gene, the 3′ end of the enhancer is required for long-range regulation and full expression of Shh in the developing forelimbs and hind limbs and, consequently, complete digit sets. The Shh locus was further observed by 3D-FISH to loop out of its chromosome territory when the gene is active. Such looping out suggests extensive unfolding of the locus and should not be confused with chromatin looping per se, which in essence refers to proximity between distal regions.

Whether or not looping out of a CT accompanies the formation of chromatin loops during long-range regulation might actually depend on the type of enhancer mechanism used to activate transcription. Indeed, while chromatin looping has been the preferred model to explain how enhancers activate promoters, there are several variations differing in the way in which the enhancer-promoter interaction is established. Enhancers have been suggested to use either free or facilitated diffusion to reach their promoters or to use an active mechanism such as "tracking" or "oozing". The process used by enhancers to find their targets might significantly impact whether activation is accompanied by looping out of CTs, since active mechanisms like tracking will invariably transform the chromatin composition, while others involving diffusion might not.

Given that control elements are not necessarily next to each other in the linear genome, mapping of physical contacts is particularly important to define functional connectivity. It was previously thought that enhancers mostly regulate and interact with their nearest gene(s), provided that this interaction does not cross sites bound by CTCF and cohesin. A survey of enhancer-promoter contacts assessed by 5C in 1% of the human genome sequence, which identified >1,000 long-range contacts in different cell lines, indicated that this is not likely the case. It was found that long-range promoter interactions were often not blocked by CTCF and cohesin binding sites and those enhancers physically interact with the nearest gene in only 7% of cases. Interactions between promoters and sequences were most frequent 120 kb upstream of transcription start sites and were asymmetric, suggesting a directionality of chromatin looping. Chromatin interaction peaks were depicted as loop configurations, i.e., two contacting regions with the intervening sequence excluded, but that this is not the only possible interpretation, since physical contacts are captured in cell populations by the 3C technologies. Peak 3C signals may reflect only the frequent occurrence of interactions in the sample, particularly in the presence of high cell-to-cell variation. Looping conformations within TADs actually have not yet been shown by complementary methods such as FISH.

The facts that one enhancer can have more than one target gene and, conversely, that multiple enhancers can regulate a single gene further highlight the need for mapping of chromatin contacts to understand the functional connectivity of regulatory elements. The β-globin locus provides a good example of this type of complex regulatory network. The β-globin locus harbors multiple β-globin genes that are expressed sequentially and in a tissue-specific manner during development. The β-globin protein produced from this region along with α-globin together form the two subunits of hemoglobin, which transports oxygen in the blood of mammals. Not surprisingly, disruptions of the β-globin gene or its regulation are known to cause diseases such as sickle cell anemia and β-thalassemia.

The β-globin locus and its regulation have been extensively characterized and are highly conserved in human and mouse. The locus spans a region of ~60 kb on human chromosome 11 and mouse chromosome 7 and features an ~15-kb domain upstream of the genes called the "locus control region" (LCR). The LCR consists of numerous enhancers required for proper β-globin expression during development. Specific looping of the LCR and expressed β-globin genes was first demonstrated with 3C in mouse erythroid cells, where chromatin around the LCR preferentially interacted with the active genes compared to brain tissue. A closer proximity between the LCR and expressed β-globin genes was also inferred in a separate study of the locus using RNA tagging and recovery of associated proteins (RNA-TRAP) in embryonic liver cells.

These data strongly suggested that looping between the LCR and genes was important for transcription. Further analyses showed that insertion of the LCR into a gene-dense region affected transcription at distances of up to 150 kb and was frequently associated with positioning of the LCR outside its chromosome territory. Contacts between the LCR and the active β-globin genes persisted after transcription inhibition, demonstrating that they form independently of RNA Pol II binding. Looping between the LCR and its targets at the β-globin locus was later shown to require several protein complexes. Erythroid-specific transcription factors, which include GATA-1, EKLF1, and TAL1, were independently found to be required for looping at the locus. In one study, the Ldb1 protein complex composed of GATA-1, TAL1, LMO2, and Ldb1 was shown to mediate loop formation. The Ldb1 complex is known to promote the transcription of numerous erythroid genes, including Myb, where long-range interactions similar to the ones found at the β-globin locus have been identified, and might therefore promote loop formation genome-wide in erythrocytes.

Transcription factor complexes do not appear entirely responsible for the chromatin conformation at the β-globin cluster. First, the observation that the locus itself loops out of its chromosome territory prior to gene activation, possibly toward more active regions between chromosomes, suggests considerable changes in chromatin composition during activation. Also, in addition to contacts between the LCR and the active β-globin genes, a network of interactions was found to link DNase I-hypersensitive sites from both sides of the locus, the LCR, and the active gene(s). A similar 3D clustering pattern, termed an "active chromatin hub" (ACH), was observed at the active α-globin locus, suggesting that both hemoglobin components are regulated by a conserved spatial mechanism. The ACH is thought to reflect the types of *cis*-acting regulatory elements that come together in the three-dimensional space to coordinate gene expression, and the nuclear compartmentalization provided by the formation of an ACH was suggested to promote transcription irrespective of the surrounding chromatin activity. CTCF binds at hypersensitive sites on either side of the β-globin locus and is required for ACH formation. Interestingly, depletion of CTCF destabilized chromatin looping at the β-globin locus and altered its histone acetylation and methylation profiles but did not significantly affect gene expression, pointing to a predominant insulator role in ACH function.

CTCF looping is also important to control the physical access of enhancers to promoters. Studies have thus far identified many more active enhancers than promoters in the human genome, and contacts between them therefore must be tightly regulated. Such regulation might be achieved by compartmentalizing inactive genes away from enhancers by differential CTCF looping, as was described for the Apo lipoprotein locus, or by domain formation at the level of sub-TADs and TADs to insulate and alter the three-dimensional path of chromatin.

## Transcription Factories and Trans Contacts

The physical clustering of actively transcribed genes into "transcription factories" was first observed when nascent transcripts were monitored by pulse labeling in HeLa cells. These transcripts were shown to colocalize with RNA polymerase II in foci that also contained splicing and transcription factors as well as chromatin-remodeling enzymes. The number of foci observed in interphase nuclei ranges from hundreds to a few thousand and appears to depend on both the cell type and the imaging technique used to detect them. In any case, multiple genes can be predicted to share the same transcription factory given that active genes far outnumber the total foci detected at any given time in the nuclei of cells. Accordingly, active genes were shown to colocalize into factories, and the transient crowding of enzymes at these sites is thought to enhance transcription and splicing efficiency.

In some cases, gene activation has actually been found to correlate well with relocalization into transcription factories containing other active genes. Although some colocalized genes may be coregulated by the same set of transcription factors, there more often appears to be little in common between the genes other than their transcription state. For instance, work on the globin genes shows that they can localize in factories with other active but unrelated genes, but they have also been found around splicing speckles. Although focal concentrations of RNA polymerase II have been shown to occur transiently and thus may not always represent real factories, genome-wide chromatin conformation analysis with TCC supports their existence genome-wide as a cluster of active genes without a shared purpose or function.

Long-range contacts are likely to constrain how genes are organized in chromosomes and how chromosomes are positioned in the nucleus. The overall tendency of active gene-rich regions to cluster into transcription factories may thus play an important role in genome organization as a source of chromatin loops and by compartmentalizing coregulated genes. The extent to which transcription factories may function in genome architecture is unknown, but their limited number and spatial positions were previously suggested to promote self-organization into tissue-specific conformations. Regardless of the mechanisms at play, three-dimensional modeling of chromatin interaction data supports a major role for looping in genome organization because of its considerable impact on the entropy of chromatin fibers.

## New Insight from Superresolution Microscopy

Several fluorescence imaging methods have overcome the diffraction resolution limit of light. Of these, there are three main techniques (required for imaging of internal cell structures)- structured illumination microscopy (SIM), stimulated emission depletion (STED), and photoactivation localization microscopy/stochastic optical reconstruction microscopy (PALM/STORM).

While STED and PALM/STORM have been used predominantly to image large protein clusters and organelles located in/on the cell membrane or in the cytoplasm, SIM has been employed to gain increased insight into nuclear ultrastructures and interchromosomal topography. SIM illuminates a sample with a series of high-spatial-frequency stripes, recording a series of frames at different stripe orientations with different shift positions, which results in 9 to 15 frames per final superresolution image. It is a hybrid technique: the whole field of view is imaged as in standard wide-field microscopy while the sample is being scanned with the stripe pattern in the manner of confocal microscopy. By illuminating the sample using a striped pattern while rotating the orientation, fluorescing signals can be captured at different times, thus resolving structures that are closer to each other than would otherwise be permitted by the actual light diffraction limit. The lateral resolution limit is halved to ~100 nm, and indeed, the resolution is doubled in all dimensions.

At the subnuclear level, SIM was first used to resolve peripheral nuclear ultrastructures such as the nuclear pore complex and the nuclear lamina. The Cremer group combined 3D-SIM with 3D-FISH in a proof-of-principle study that showed that key chromatin features were largely well preserved (but with some perturbations) after 3D-FISH down to the resolution limit imposed by 3D-SIM. This combined technique, in conjunction with 5C, revealed that adjacent DNA sequences within the same TAD colocalized to a greater extent than did adjacent sequences in different TADs. Also, the combination of 3D-SIM and 3D-FISH showed that chromatin decompaction occurs at key differentiation genes and that these genes migrate to the center of the nucleus as neural differentiation progresses during embryonic development. 3D-SIM was also used to show a striking difference in the functional organizations of transcriptionally active CTs and the Barr body. In a novel approach based on SIM but replacing the single light beam with a lattice light sheet that enabled single-molecule live imaging, Sox2 binding sites were mapped and shown to form discrete clusters in live ES cells.

The three techniques have strengths and weaknesses. The decision of which one to use depends on several parameters such as resolution and whether the specimen is fixed or live imaging is to be done. So far, only SIM has been combined with FISH to investigate chromatin ultrastructure in reported studies; however, the other techniques should also be able to provide further insight into chromatin topography, particularly STED microscopy when looking at interprobe distances within compact domains.

## Key Regulators of Genome Architecture

Despite the fact that radial chromosome positions can vary significantly between generations, whether chromatin organization is itself epigenetic at high resolution is unknown, although it was recently suggested that by mutually affecting each other, the chromatin state and architecture take part in a self-enforcing feedback process to propagate cell fate memory. Proteins that can both physically shape and regulate the composition of chromatin are thus likely to play important roles in spatial inheritance. The CTCF protein and the cohesion complex are two chromatin components thought to shape the human genome in hierarchical length scales, which have been linked to transcription regulation, imprinting, and X chromosome inactivation.

## CTCF as a Master Genome Organizer

The CCCTC-binding factor (CTCF) is an essential protein that is highly conserved from fly to human. It is known to exert vastly diverse nuclear functions, and its functional diversity is thought to originate from the way in which it binds DNA. CTCF binds genomic DNA through a central 11-zinc-finger DNA binding domain with close to 100% homology between chicken, mouse, and human. It can bind to a wide range of sequences by the combinatorial use of its zinc fingers, but most binding sites (75 to 90%) contain a core consensus of 11 to 15 bp. It was postulated that both CTCF and DNA adopt different conformations upon binding to accommodate different zinc finger combinations based on the underlying sequence and that these allosteric shifts determine the kinds of proteins that can bind CTCF. CTCF is thus viewed as a "multivalent factor" because it binds to different proteins depending on the sequence with which it interacts, leading to different posttranslational modifications of itself and surrounding proteins and different functional outcomes.

CTCF is a vertebrate protein shown to bind insulator sequences directly and to help establish their activity. Insulators are DNA elements that control transcription either by stopping the spread of histone marks or by preventing contacts that activate transcription. CTCF is known to regulate gene expression by both mechanisms. For instance, it can act as an insulator/barrier at heterochromatin boundaries and divide chromatin into silent and active domains. It is also known for its enhancer-blocking function to repress transcription, an activity that likely plays a pivotal role in promoting or preventing long-distance contacts between regulatory elements and target genes.

Highlighting its importance in the regulation of gene expression, CTCF was found to bind >30,000 sites across the human genome, and many of these sites are conserved across cell types and species. Genome-wide analyses showed that groups of genes within regions flanked by CTCF are likely to be coregulated, in contrast to gene pairs divided by CTCF binding, a characteristic that may be linked to CTCF's ability to demarcate chromatin domains. Regions that frequently bind CTCF often contain genes featuring different tissue-specific promoters, suggesting that it might be involved in the complex regulation of such genes. CTCF also appears to be involved in the formation of lamina-associated domains, as suggested by its enrichment at LAD boundaries and by ChIA-PET interaction data. Together, these results point to a general role for CTCF in genome function, partly by segregating transcriptional activity to specific nuclear areas.

The exact mechanism(s) by which CTCF contributes to insulator function is unknown, but evidence strongly suggests that it involves the manipulation of chromatin architecture. Indeed, CTCF was shown to mediate long-range chromatin interactions such as those observed during enhancer-promoter looping. In fact, CTCF's roles in transcription, imprinting, and X chromosome inactivation could likely be explained mainly by its ability to form long-range DNA contacts and spatially organize the chromatin. In addition, CTCF was found to mediate contacts both within and between chromosomes. One of the mechanisms by which CTCF might physically recruit remote sites along and between chromosomes is through its ability to oligomerize. Exactly how the protein achieves this is unknown, but CTCF was found to bind asymmetrically across strong topological borders in a manner that predicts the directionality of CTCF-CTCF interactions, suggesting that the types of long-range contacts made by the protein are defined by the position and orientation of binding sites along the linear sequence.

One of the most well-characterized CTCF chromatin architectures was identified at the imprinted Igf2/H19 locus and regulates imprinting. It was found that Igf2 repression on the maternal allele is achieved by preventing the interaction between the gene and a distal enhancer through the formation of chromatin loops mediated by CTCF. In contrast, CTCF binding at the ICR and insulator looping are prevented by DNA methylation on paternal alleles, allowing the Igf2 gene to contact the distal enhancer by transcription factor-mediated looping. These studies were the first ones to demonstrate cross talk between "classical" epigenetics and spatial chromatin organization.

CTCF looping is essential for gene regulation and relevant to human health. For example, mutations that affect CTCF binding at the H19/Igf2 locus were shown to result in serious human syndromes. Improper CTCF binding has also been linked to other diseases, such as Huntington's disease, where mutations that destabilize CTCF binding sites appear to cause trinucleotide repeat expansion.

## Cohesin as a Cell Type-specific Regulator of Chromatin Organization

Cohesin is another important genome organizer involved in transcription regulation. Cohesin is a multisubunit protein complex composed of the Smc1A, Smc3, Rad21, and Stag1/2 (SA1/2) proteins. It was initially recognized for its role in sister chromatid cohesion, mitotic and meiotic chromosome segregation, and DNA repair. The first indication that cohesin regulates transcription was the finding that mutations in Nipped-B facilitate the activation of *Drosophila cut* and *Utrabithorax* homeobox genes by distal transcriptional enhancers. Nipped-B and its human orthologue Nipped-B-like (NIPBL) are factors required for the loading of cohesin onto the DNA that colocalize with CTCF/cohesin but also bind at independent sites like promoters. Mutations in the cohesin Smc1A or Smc3 subunit and in the NIPBL gene are responsible for many cases of Cornelia de Lange syndrome (CdLS). Like CTCF, cohesin was also found to bind thousands of sites in interphase nuclei, but its binding is much more tissue specific. Cohesin was shown to colocalize with Mediator genome-wide and facilitated enhancer-promoter looping at the *Nanog* gene. Although it was originally thought to be important for mouse stem cell maintenance by directly regulating the *Oct4* and *Nanog* pluripotent genes, a more recent study indicates that this is not likely the case.

Cohesin colocalizes extensively with CTCF throughout the genome. In fact, many of the original CTCF-mediated looping contacts were later found to require cohesin. However, CTCF does not exclusively colocalize with cohesin and vice versa. Also, sites bound by CTCF and cohesin, colocalized or not, can each colocalize with Mediator. Together, these findings point to the existence of functionally distinct CTCF and cohesin complexes that are DNA bound and involved in defining the chromatin architecture. How distinct CTCF/cohesin complexes relate to each other and genes to coordinate architecture and transcription is unknown. However, the fact that CTCF and cohesin were found to be enriched at TAD boundaries suggests that they at least play a role in partitioning the transcriptional landscape of the genome.

## Roles of CTCF and Cohesin in TAD Formation

The enrichment of CTCF and cohesin at TAD boundaries is one of the most interesting TAD features and the subject of much scrutiny. While clearly enriched at boundaries, the absolute number of CTCF and cohesin binding sites within the TADs themselves is much greater suggesting multiple functions for the proteins and/or that TAD structures may be more complex than currently thought. CTCF is nonetheless required for proper TAD formation since its depletion results in fewer intra-TAD contacts and in more inter-TAD interactions.

If CTCF truly contributes to delineating TADs, it might do so by mechanisms similar to the ones used at heterochromatin boundary sites. Supporting this possibility is one study where deletion of a TAD boundary led to the formation of contacts across the deleted region and transcription misregulation. The interaction profiles of TADs point to higher enhancer-promoter interaction frequencies within domains than between them. Thus, by physically segregating chromatin regions into topological domains, CTCF and cohesin might define functional microenvironments for regulatory elements and target genes where contacts are more easily nucleated while preventing chromatin states from spreading and limiting contacts with the rest of the genome. This model is supported by the fact that partitioning of the genome into TADs correlates with enhancer-promoter units, clusters of coregulated promoters and enhancers. Physical modeling of 3C-type data that explored all possible TAD conformations within population-averaged data sets further suggested that contacts between control elements dynamically fluctuate rather than exist as stable structures.

The role of cohesin at TAD boundaries and its relationship with CTCF are unclear. In contrast to CTCF, cohesin depletion only reduces the intensity of intra-TAD interactions without affecting the actual TAD location or organization, which is consistent with a role for cohesin in mediating tissue-specific enhancer-promoter contacts at the submegabase scale. Cohesin is known to be required for CTCF-based insulation, and depletion affects insulation patterns genome-wide, correlates with global gene expression changes, and negatively affects hierarchical long-range interactions between TAD boundaries separated by multiple domains. These findings collectively suggest that cohesin may organize chromatin in such a way as to prevent interactions between particular TADs and isolate gene expression states from one another. The mechanism by which cohesin exerts this regulation on chromatin organization remains unclear but might depend on CTCF.

# DNA Sequencing

DNA sequencing is a method used to determine the precise order of the four nucleotide bases – adenine, guanine, cytosine and thymine - that make up a strand of DNA. These bases provide the underlying genetic basis (the genotype) for telling a cell what to do, where to go and what kind of cell to become (the phenotype). Nucleotides are not the only determinants of phenotypes, but are essential to their formation. Each individual and organism has a specific nucleotide base sequence.

The two scientists in the photograph are reading the genetic code for a DNA sample on a highlighted light board. Such analysis is usually done by a computer.

## Importance

DNA sequencing played a pivotal role in mapping out the human genome, completed in 2003, and is an essential tool for many basic and applied research applications today. It has for example provided an important tool for determining the thousands of nucleotide variations associated with specific genetic diseases, like Huntington's, which may help to better understand these diseases and advance treatment.

DNA sequencing also underpins pharmacogenomics. This is a relatively new field which is leading the way to more personalised medicine. Pharmacogenomics looks at how a person's individual genome variations affect their response to a drug. Such data is being used to determine which drug gives the best outcome in particular patients. Over 140 drugs approved by the FDA now include pharmacogenomics information in their labelling. Such labelling is not only important in terms of matching patients to their most appropriate drug, but also for working out what their drug dose should be and their level of risk in terms of adverse events. Individual genetic profiling is already being used routinely to prescribe therapies for patients with HIV, breast cancer, lymphoblastic leukaemia and colon cancer and in the future will be used to tailor treatments for cardiovascular disease, cancer, asthma, Alzheimer's disease and depression. Drug developers are also using pharmacogenomics data to design drugs which can be targeted at subgroups of patients with specific genetic profiles.

DNA sequencing provides the means to know how nucleotide bases are arranged in a piece of DNA. Several methods have been developed for this process. These have four key steps. In the first instance DNA is removed from the cell. This can be done either mechanically or chemically. The second phase involves breaking up the DNA and inserting its pieces into vectors, cells that indefinitely self-replicate, for cloning. In the third phase the DNA clones are placed with a dye-labelled primer (a short stretch of DNA that promotes replication) into a thermal cycler, a machine which automatically raises and lowers the temperature to catalyse replication. The final phase consists of electrophoresis, whereby the DNA segments are placed in a gel and subjected to an electrical current which moves them. Originally the gel was placed on a slab, but today it is inserted into a very thin glass tube known as a capillary. When subjected to an electrical current the smaller nucleotides in the DNA move faster than the larger ones. Electrophoresis thus helps sort out the DNA fragments by their size. The different nucleotide bases in the DNA fragments are identified by their dyes which are activated when they pass through a laser beam. All the information is fed into a computer and the DNA sequence displayed on a screen for analysis.

The method developed by Sanger was pivotal to the international Human Genome Project. Costing over US$3 billion and taking 13 years to complete, this project provided the first complete Human DNA sequence in 2003. Data from the project provided the first means to map out the genetic mutations that underlie specific genetic diseases. It also opened up a path to more personalised medicine, enabling scientists to examine the extent to which a patient's response to a drug is determined by their genetic profile. The genetic profile of a patient's tumour, for example, can now be used to work out what is the most effective treatment for an individual. It is also hoped that in the future that knowing the sequence of a person's genome will help work out a person's predisposition to certain diseases, such as heart disease, cancer and type II diabetes, which could pave the way to better preventative care.

Data from the Human Genome Project has also helped fuel the development of gene therapy, a type of treatment designed to replace defective genes in certain genetic disorders. In addition, it has provided a means to design drugs that can target specific genes that cause disease.

Beyond medicine, DNA sequencing is now used for genetic testing for paternity and other family relationships. It also helps identify crime suspects and victims involved in catastrophes. The technique is also vital to detecting bacteria and other organisms that may pollute air, water, soil and food. In addition the method is important to the study of the evolution of different population groups and their migratory patterns as well as determining pedigree for seed or livestock.

Since the completion of the Human Genome Project, technological improvements and automation have increased speed and lowered costs to the point where individual genes can be sequenced routinely, and some labs can sequence well over 100,000 billion bases per year, and an entire genome can be sequenced for just a few thousand dollars.

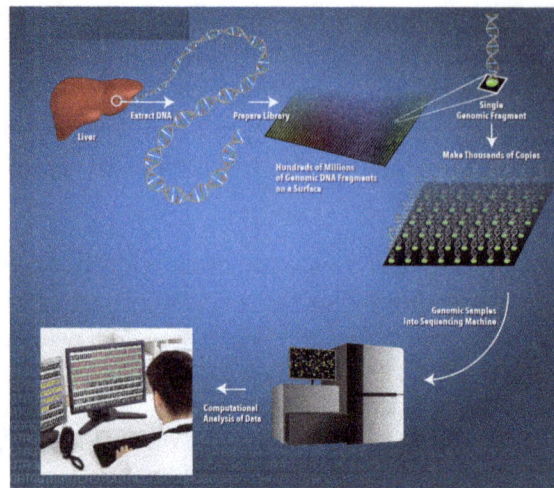

Many of these new technologies were developed with support from the National Human Genome Research Institute (NHGRI) Genome Technology Program and its Advanced DNA Sequencing Technology awards. One of NHGRI's goals is to promote new technologies that could eventually reduce the cost of sequencing a human genome of even higher quality than is possible today and for less than $1,000.

Since the completion of the Human Genome Project, technological improvements and automation have increased speed and lowered costs to the point where individual genes can be sequenced routinely, and some labs can sequence well over 100,000 billion bases per year, and an entire genome can be sequenced for just a few thousand dollars. Many of these new technologies were developed with support from the National Human Genome Research Institute (NHGRI) Genome Technology Program and its Advanced DNA Sequencing Technology awards. One of NHGRI's goals is to promote new technologies that could eventually reduce the cost of sequencing a human genome of even higher quality than is possible today and for less than $1,000.

One new sequencing technology involves watching DNA polymerase molecules as they copy DNA - the same molecules that make new copies of DNA in our cells - with a very fast movie camera and microscope, and incorporating different colors of bright dyes, one each for the letters A, T, C and G. This method provides different and very valuable information than what's provided by the instrument systems that are in most common use.

Another new technology in development entails the use of nanopores to sequence DNA. Nanopore-based DNA sequencing involves threading single DNA strands through extremely tiny pores in a membrane. DNA bases are read one at a time as they squeeze through the nanopore. The bases are identified by measuring differences in their effect on ions and electrical current flowing through the pore. Using nanopores to sequence DNA offers many potential advantages over current methods. The goal is for sequencing to cost less and be done faster. Unlike sequencing methods currently in use, nanopore DNA sequencing means researchers can study the same molecule over and over again.

Researchers now are able to compare large stretches of DNA - 1 million bases or more - from different individuals quickly and cheaply. Such comparisons can yield an enormous amount of information about the role of inheritance in susceptibility to disease and in response to environmental influences. In addition, the ability to sequence the genome more rapidly and cost-effectively creates vast potential for diagnostics and therapies.

Although routine DNA sequencing in the doctor's office is still many years away, some large medical centers have begun to use sequencing to detect and treat some diseases. In cancer, for example, physicians are increasingly able to use sequence data to identify the particular type of cancer a patient has. This enables the physician to make better choices for treatments.

Researchers in the NHGRI-supported Undiagnosed Diseases Program use DNA sequencing to try to identify the genetic causes of rare diseases. Other researchers are studying its use in screening newborns for disease and disease risk.

Moreover, The Cancer Genome Atlas project, which is supported by NHGRI and the National Cancer Institute, is using DNA sequencing to unravel the genomic details of some 30 cancer types. Another National Institutes of Health program examines how gene activity is controlled in different tissues and the role of gene regulation in disease. On-going and planned large-scale projects use DNA sequencing to examine the development of common and complex diseases, such as heart disease and diabetes, and in inherited diseases that cause physical malformations, developmental delay and metabolic diseases.

Comparing the genome sequences of different types of animals and organisms, such as chimpanzees and yeast, can also provide insights into the biology of development and evolution.

# DNA Sequencing Technologies

The development of a draft human genome sequence, which has subsequently been revised to constitute a "reference" human genome sequence, facilitated the development of "next generation" sequencing (NGS). NGS is a broad term that refers to a set of methods for: 1) genomic template preparation, or the methodology for processing genomic DNA for downstream sequencing; 2) near simultaneous, or "massively-parallel", generation of millions to billions of short sequence reads; 3) alignment of sequence reads to a reference sequence; 4) sequence assembly from aligned sequence reads and genetic variant discovery. Most investigators use the output from this final step, a list of genotypes for positions with at least one allele that differs from a reference sequence ("variants") in all downstream analysis.

Thus, "whole genome sequence" data generally refers not to ~3 billion diploid genotypes that cover the known chromosomal positions but the 3–4 million genotypes in each genome that differ from the reference sequence. Several NGS technologies exist that differ primarily in methods for clonal amplification of short fragments of DNA and sequencing the resulting short DNA fragments. Each has specific advantages in terms of read length, accuracy, and throughput. All currently forego the time-consuming bacterial cloning step that was used for library preparation in the Human Genome Project. One issue that deserves specific mention is that of read length. Shorter sequence reads (100 base pairs or shorter) are well suited to the biochemical reactions employed by most of the sequencing technologies. However, the generation of short reads complicates sequence assembly, particularly in repetitive regions of the genome. The generation of longer sequence reads (1000 base pairs or longer) simplifies this task. Furthermore, the use of longer sequence reads spanning several variants aids in resolution of "haplotype phase", which is the assignment of each allele in a heterozygous genotype to one chromosome of each homologous pair, e.g., the assignment of an "A" allele in a "A/G" genotype to a paternally-derived segment of chromosome.

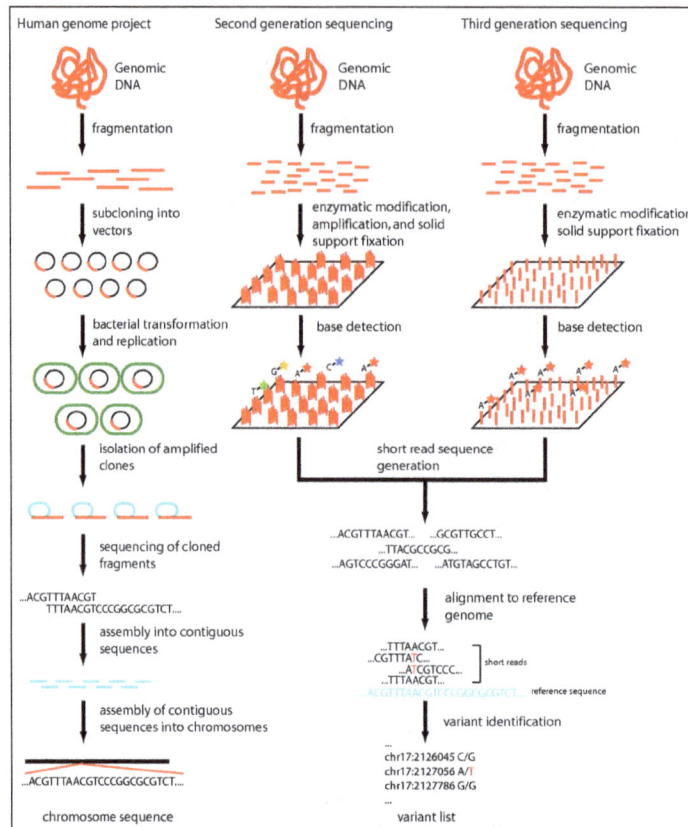

Figure: Three generations of human genome sequencing technology.

Three groups of sequencing technology are depicted: sequencing in the human genome project; second generation sequencing as exemplified by the Illumina HiSeq 2000; third generation sequencing as exemplified by the Helicos Heliscope single molecule sequencer.

Table: Sequencing Platform Comparison.

| Platform | Amplification | Sequencing | Detection | Read Length | Output per Run | Run Time |
|---|---|---|---|---|---|---|
| Second-generation sequencing platforms | | | | | | |
| 454 | Emulsion PCR on beads | Unlabeled nucleotide incorporation | Detection of light emitted by release of $PP_i$ | Variable (400 bp for single end sequencing) | 400-600 Mbp | 10 h |
| SOLiD | Emulsion PCR on beads | Ligation of 2- base encoded fluorescent oligonucleotides | Fluorescence emission from labelled oligonucleotides | 75+35 bp | 20-30 Gbp | 7 d |
| Illumina | Array-based enzymatic amplification | Fluorescently labeled end-blocked nucleotide incorporation | Fluorescence emission from nucleotides | 2[times]10 bp | 100-200 Gbp | 8 d |
| Complete | Rolling-circle replication of short segments of DNA into nanoballs | Ligation of fluorescently labelled oligonucleotide probes | Fluorescence emission from oligonucleotide probes | 2[times]35 bp | 20-60 Gbp | 12 d |

| Third-generation sequencing platforms | | | | | | |
|---|---|---|---|---|---|---|
| Helicos | NA | Single dye-labelled nucleotides are added sequentially and incorporated by polymerases by use of single DNA molecular templates | Microscopy of fluorescently labelled nucleotides | 2[times]25-55 bp | 21-35 Gbp | 8 d |
| Pacific Biosciences | NA | Incorporation of fluorescently labelled nucleotides by polymerases on solid support | Zero-mode waveguide imaging of fluorescent nucleotide incorporation by individual polymerases | 2[times]1000 bp | 75-100 Mbp (projected); 5-10 Mbp (actual usable sequence) | 30 min |
| Oxford nanopore | NA | Processive endo- or exonuclease activity feeds individual bases or whole DNA strands through protein or solid-state nanopores | Current disruption across nanopore corresponds to nucleotide structure | Variable | Variable | Variable |
| Ion Torrent | Variable | DNA polymerase incorporation of unlabelled nucleotides added sequentially to solid-state microwells | Solid-state detection of hydrogen ions released by nucleotide incorporation | 200 bp | 10 Mbp to 1 Gbp | 2 h |

Where: bp indicates base pair; Gbp, one billion base pairs; Mbp, one million base pairs; $PP_i$, pyrophosphate; and PCR, polymerase chain reaction.

Of the NGS platforms that are currently commercially available, the 454 (454 Life Sciences/Roche) instrument was developed first. This platform is based on "pyro sequencing" which detects light emitted by secondary reactions initiated by the release of pyrophosphate during nucleotide incorporation. Advantages include long reads and facile "mate-pair" sequencing, a method that sequences both ends of a previously circularized DNA molecule. Pairing reads that span tens of kilobases of genomic template sequence further facilitates haplotype phasing and the identification of structural genetic variation such as deletions and insertions of large segments of DNA. Disadvantages include systematic errors in reading frame ("frame shift errors") in certain circumstances and lower throughput and higher sequencing costs than other commercial technologies.

Solid (Applied Biosystems by Life Technologies) sequencing utilizes sequencing-by-ligation in which the sequence of a DNA template is read by competitive ligation of 2-base probes to the nascent DNA strand. Advantages include throughput (~20–30 Gbp per run), and base-level error information encoded in the 2-base sequences, both of which make the platform suitable for human whole genome and exome variant discovery. The main disadvantage is the necessity to work with unconventional data formats for sequence reads and the reference genome.

The Illumina/Solexa platform is widely used for a variety of applications, including human whole genome and exome variant discovery and transcriptome sequencing ("RNAseq"), by virtue of easily prepared paired-end sequencing libraries, high throughput, and ease of analysis of its short read information. After genomic DNA isolation, fragmentation, and several enzymatic modification steps, sequencing libraries are amplified from single DNA strands on glass surfaces. The resultant templates are sequenced using an approach in which fluorescently labeled "end-blocked nucleotides," which do not allow further DNA polymerization, are incorporated by DNA polymerase, the base-specific fluorescent color is detected via fluorescence imaging, the end block and fluorescent tag is enzymatically cleaved, and the process is repeated following image storage, yielding image-encoded nucleotide sequences. Drawbacks include comparatively short sequence reads ($\leq 100$ bp) and practical limits to insert sizes for paired end sequencing.

Complete Genomics, Inc. provides a sequencing service, in contrast to other companies that have primarily focused on providing sequencing instruments, that is targeted solely towards human whole genomes. The instrument uses sequencing-by-ligation of hundreds of "DNA nano-balls," or chained-replicates of 70-base-pair sequences of sheared genomic DNA modified by adaptor inserts. Theoretical throughput exceeds that of any of the NGS technologies described thus far.

## Third Generation Sequencing Technologies

A "third" generation of sequencing instruments has been developed that is defined by the lack of DNA or RNA amplification in template library preparation ("single molecule sequencing"). By foregoing this step, these technologies require less genomic DNA, avoid PCR-introduced error and amplification bias, and may be superior for high-throughput sequencing applications, such as transcriptome sequencing ("RNAseq"), that depend on accurate quantification of relative DNA or RNA fragment abundance.

The first of these single-molecule sequencing technologies is the Helicos Heliscope (Helicos Bio Sciences). The specific Helicos chemistry is based on single-molecule sequencing by cyclic reversible terminator nucleotide incorporation. A single dye molecule is used to label the dNTPs and fluorescence microscopy is used to image the dye in sequencing reactions carried out on single molecule templates on solid support. The order in which each fluorescently labeled dNTP is added to the sequencing reactor determines the base sequence at that position. Notably, the instrument is also suitable for direct RNA sequencing without conversion to complementary DNA (cDNA), thus avoiding error and copy number bias associated with reverse transcription.

Pacific Biosciences have recently developed a method for imaging individual DNA polymerase molecules as they synthesize a nascent DNA molecule covalently attached to solid support. Advantages include read information that is theoretically as long as 1 kb or longer and real-time sequencing kinetics that reflect nucleotide methylation state and DNA secondary structure.

Life Technology's Ion Torrent device is targeted towards individual laboratories interested in a small footprint, medium throughput sequencing platform. This sequencing engine is based on detection of hydrogen ions released from nucleotides incorporated into the growing DNA strand. This signal is detected in a solid-state semiconductor akin to a miniaturized pH meter, and the technology is theoretically suitable to single-molecule sequencing. Throughput is currently low (< 1 Gbp per run), but the release of higher density chips has made sequencing of transcriptomes and exomes feasible.

"Nanopore" sequencing technologies detect base-specific changes in ionic flux as DNA traverses small pores in solid surfaces that are placed in an electric field. Advantages to this method include theoretically unparalleled sequencing speed and minimal template preparation. At this point, however, detection speed and accuracy remain significant technological hurdles, as the transit speed of nucleic acids through nanopores in even minimal electric fields is several orders or magnitude higher than the highest detection frequency. Several enzymatic methods have been developed to slow transit time and facilitate detection of changes in ionic flux.

## Processing High-throughput Sequence Data

Data generation from high-throughput sequencing is becoming less expensive and time consuming. Generating sequence data, however, is only the first step in extracting usable information from high-throughput sequencing. For output from most currently available sequencing platforms, several tasks must be performed prior to downstream analysis: 1) short read mapping, or alignment of each sequence read to a reference genome to identify the genomic sequence represented by the short read; 2) base calling at every genomic position covered by aligned short reads; 3) identification of sequence variation from the reference genome. The percentage of base positions that are read by properly aligned short reads is described by "coverage." The number of times that a single base position is read by short read sequences is termed "depth of coverage" and most investigators currently consider 30-fold ("30 times") average depth of coverage as a benchmark for high-quality genome sequence data.

## Aligning Sequence Reads to the Human Reference Genome

There are several programs for mapping short reads to a reference genome; for an in-depth comparison of alignment programs, we direct the reader to a recent work by Li and Homer. Historically, mapping alignment with quality ("MAQ") was the most widely used alignment algorithm, but this algorithm has been supplanted by other open-source solutions that are superior for longer (>35 bp) sequence reads. Though several alignment algorithms can be run on high-memory multiple core desktops and even laptops, parallel computing architecture, which utilizes multiple processors to perform alignment tasks simultaneously, reduces the time required for alignment several fold. Unfortunately, few individual labs currently are able to provide this computing power. One solution is on-demand distributed or parallel computing architecture, i.e., "cloud" computing. This approach is economical in the sense that elastic parallel computing environments allow users to select and utilize only processing and storage capacity necessary for current tasks.

## Identifying Single Nucleotide Variants and Small Insertions/Deletions

Following alignment to the reference genome, sequence reads are compared at every genomic position, producing a base call for each chromosomal position. A variety of different algorithms incorporate base quality, which specifies the confidence of each base call within the individual short reads, mapping quality, or confidence of accurate mapping of each short read to the specified genomic locus, and the number of bases contributing to each of the possible 16 genotypes at a position, into a probabilistic score for genotypes at every chromosomal location. The most likely genotype is compared to the reference sequence, and, typically, only positions containing at least one base differing from the reference sequence are retained for downstream analysis. This fact has several important implications. First, the reference base is crucial to the identification of genetic variation: if the haploid reference base harbors the same allele predisposing to disease as the subject being sequenced, it will not appear in the variant list, potentially leading to underestimation of the burden of certain disease-associated alleles. Second, comparison between individuals, e.g., in co-segregation and linkage studies, can be complicated by the degree of overlap between genetic variant sets such that the assumption of homozygous reference allele calls can bias exploratory studies for causative variants. Several variant calling solutions, notably, SAM tools and the Genome Analysis Toolkit (GATK) have base calling algorithms that facilitate cohort-wide variant identification, which addresses this problem. Third, the reference sequence represents a small sampling of human genetic variation, and as large scale sequencing efforts are undertaken, ethnicity-specific major allele differences may impact alignment of short reads against the current reference genome and subsequent variant identification.

## Identifying Large Structural Variants

Large structural rearrangements >1kb, termed structural variants (SVs), encompass large deletions, duplications, insertions, and inversions, and transposons. Largely ignored in many early sequencing efforts, emerging evidence suggests that these structural variants are strongly associated with several Mendelian and complex diseases, including familial dilated cardiomyopathy, autism spectrum disorders, idiopathic mental retardation, schizophrenia, and Crohn's disease. In some cases these large genetic variants underly > 15% of disease diagnoses. Several methods have been developed for identification of SVs, but three main methods have generally been accepted and are used for identification of specific types of SVs. Notably, however, due to high false positive rates for SV detection using high-throughput sequencing, this and other PCR-based methods are often used to confirm candidate SVs.

The first method for identification of structural variants is mate pair sequencing, which is based on sequencing two ends of a DNA molecule following circularization, providing paired short read sequence information separated by hundreds to thousands of base pairs. A related technique, paired end sequencing, is used routinely in most commercial sequencing technologies to provide paired short sequence reads from each end of an amplified linear DNA molecule. Comparison of median insert size and orientation from paired end reads to homologous chromosomal segments in the reference genome is used to identify structural rearrangements. Though sensitive for inversions and other "copy neutral" SVs, or SVs that do not change the copy number of the affected chromosomal region, and somewhat well suited to identifying start and end points of SVs ("breakpoints"), detection scope is limited by the size of the insert, in that only structural rearrangements spanned by the insert can be detected.

A second method for identification of structural rearrangements is based on regional variation in read depth, which is in turn dependent on copy number of the genomic region interrogated. Several methods have been developed for identification of significant differences in read depth in genomic regions relative to median read depth. This method for identification of SVs is ideally suited for identification of large insertions and deletions, but has limited capability to resolve breakpoints, and cannot distinguish copy neutral SVs from normal sequence.

The third method for identification of large SVs is split-read mapping, which is based on mapping elements with inserts in the reference genome or the sample genome to contiguous short read sequences by using one end of the read as an anchor and the other end to search for possible breakpoints, yielding single-nucleotide level breakpoint resolution and novel sequence discovery in some cases. Finally, candidate structural variants are often compared to known structural variants identified using population-scale sequencing or genotyping to provide probabilities of false discovery and improved breakpoint resolution.

## Variant Quality Control and Genotype Validation

Validation of sequence data has become a particularly difficult problem in interpretation of genetic variants discovered via high-throughput sequencing. Per genotype error rates for commercially available high-throughput sequencing technologies achieving an average depth of coverage of >30× are currently between one in every 1000 to one in every 100,000 bases. By comparison, per-genotype error rates for Sanger sequencing, the current standard for clinical applications, is between one in 100,000 and one in 1,000,000 base pairs. Filtering variants via a combination of quality score metrics for individual short reads and final genotypes can minimize errors. Roach, et al, have demonstrated that leveraging family genotype information can also be useful for error identification, in that pedigree-based allele inheritance analysis can be used to identify not only inconsistencies with Mendel's laws of inheritance, but regions in which short reads have been incorrectly mapped or genotyped. We have recently demonstrated a >90% reduction in the error rate by sequestering variants identified in these regions.

Despite these and other advances in error reduction, however, high-throughput sequencing platforms do not yet provide the level of confidence about individual variants that would be required for routine incorporation into clinical care. To date, clinically important variants have mostly been re-sequenced using Sanger-based chemistry or confirmed with oligonucleotide genotyping arrays. Both approaches are time and resource-intensive. Alternative capture-based approaches, in which either a standard commercial or custom oligonucleotide set is used to select genomic regions of interest for high-coverage high-throughput re-sequencing, are also costly and time-consuming. Validation of small structural variants such as insertions and deletions is even more difficult, often requiring bacterial cloning of single strands prior to re-sequencing. Until the accuracy of high-throughput sequencing improves such that primary data does not require orthogonal confirmation, data validation will continue to be a major barrier to widespread incorporation of high-throughput sequence data into clinical applications.

The first DNA sequencing method devised by Sanger and Coulson in 1975 was called plus and minus sequencing that utilized E. coli DNA pol I and DNA polymerase from bacteriophage T4 with different limiting triphosphates. This technique had a low efficiency. Sanger and co-worker (1977) eventually invented a new method for DNA sequencing via enzymatic polymerization that basically revolutionized DNA sequencing technology.

The most popular method for doing this is called the dideoxy method or Sanger method (named after its inventor, Frederick Sanger, who was awarded the 1980 Nobel prize in chemistry [his second] for this achievement). Finding a single gene amid the vast stretches of DNA that make up the human genome – three billion base-pairs' worth – requires a set of powerful tools. These tools include genetic maps, physical maps and DNA sequence which are a detailed description of the order of the chemical building blocks, or bases, in a given stretch of DNA.

Scientists need to know the sequence of bases because it tells them the kind of genetic information that is carried in a particular segment of DNA. For example, they can use sequence information to determine which stretches of DNA contain genes, as well as to analyze those genes for changes in sequence, called mutations, that may cause disease.

The first methods for sequencing DNA were developed in the mid-1970s. At that time, scientists could sequence only a few base pairs per year, not nearly enough to sequence a single gene, much less the entire human genome. By the time the HGP began in 1990, only a few laboratories had managed to sequence a mere 100,000 bases, and the cost of sequencing remained very high. Since then, technological improvements and automation have increased speed and lowered cost to the point where individual genes can be sequenced routinely, and some labs can sequence well over 100 million bases per year.

DNA is synthesized from four deoxynucleotide triphosphates. The top formula shows one of them: deoxythymidine triphosphate (dTTP). Each new nucleotide is added to the 3′ − OH group of the last nucleotide added.

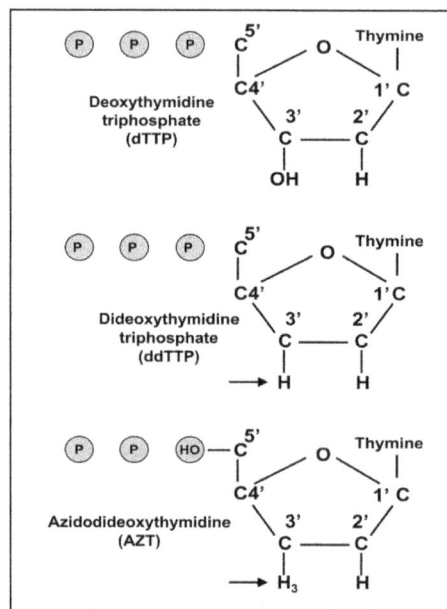

Structure of dideoxynucleotide.

The dideoxy method gets its name from the critical role played by synthetic nucleotides that lack the -OH at the 3′ carbon atom. A dideoxynucleotide (dideoxythymidine triphosphate − ddTTP) can be added to the growing DNA strand. When it is added it stops chain elongation because there is no 3′ -OH for the next nucleotide to be attached. For this reason, the dideoxy method is also called the chain termination method.

The bottom formula shows the structure of azidothymidine (AZT), a drug used to treat AIDS. AZT (which is also called zidovudine) is taken up by cells where it is converted into the triphosphate. The reverse transcriptase of the human immunodeficiency virus (HIV) prefers AZT triphosphate to the normal nucleotide (dTTP). Because AZT has no 3′ -OH group, DNA synthesis by reverse transcriptase halts when AZT triphosphate is incorporated in the growing DNA strand. Fortunately, the DNA polymerases of the host cell prefer dTTP, so side effects from the drug are not as severe as might have been predicted.

## Procedure

The DNA to be sequenced is prepared as a single strand.

This template DNA is mixed with the following:

A mixture of all four normal deoxynucleotides in sample quantities:

1. dATP

2. dGTP

3. dCTP

4. dTTP

A mixture of all four dideoxynucleotides, each present in limiting quantities and each labeled with a "tag" that fluoresces a different colour:

1. ddATP

2. ddGTP

3. ddCTP

4. ddTTP

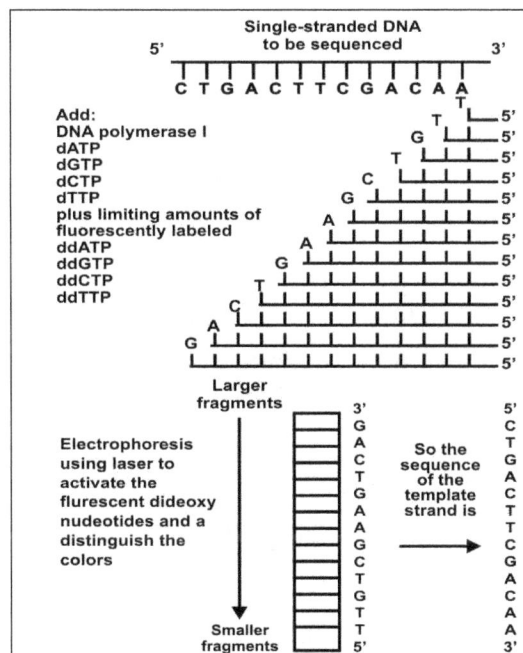

Sanger's method of DNA.

DNA polymerase I- Because all four normal nucleotides are present, chain elongation proceeds normally until, by chance, DNA polymerase inserts a dideoxy nucleotide instead of the normal deoxynucleotide. If the ratio of normal nucleotide to the dideoxy versions is high enough, some DNA strands will succeed in adding several hundred nucleotides before insertion of the dideoxy version halts the process.

At the end of the incubation period, the fragments are separated by length from longest to shortest. The resolution is so good that a difference of one nucleotide is enough to separate that strand from the next shorter and next longer strand. Each of the four dideoxynucleotides fluoresces a different colour when illuminated by a laser beam and an automatic scanner provides a printout of the sequence.

## Maxam and Gilbert Method

In 1977, Maxam and Gilbert described a sequencing method based on chemical degradation at specific locations of the DNA molecule. The end labeled DNA fragments are subjected to random cleavage at adenine, cytosine, guanine or thymine positions using specific chemical agents and the products of these fours reactions are separated using polyacrylamide gel electrophoresis (PAGE). As in Sanger method, the sequence can be easily read from four parallel lanes in the sequencing gel.

Double stranded or single stranded DNA from chromosomal DNA can be used as template. Originally, end labeling was done with P phosphate or with a nucleotide linked to P and enzymatically incorporated into the end fragment. The read length is up to 500bp. The chemical reactions in the technique are slow and involved hazardous chemicals that require special handling in the DNA cleavage reaction.

As in Sanger's method, additional cautions in Maxam and Gilbert method include purification and separation of DNA fragments and higher analysis time. Therefore, this technology is not suitable for high throughput large-scale investigation.

## Hybridization Method

Ed Southern's (1990) sequencing by hybridization technique relies on detection of specific DNA sequences using hybridization of complementary probes. It utilizes a large number of short nested oligonucleotides immobilized on a solid support to which the labeled sequencing template is hybridized. The target sequence is deduced by computer analysis of the hybridization pattern of the sample DNA.

DNA sequence can also be analyzed by sequencing by synthesis. Sequencing by hybridization makes use of a universal DNA microarray, which harbors all nucleotides of length k (called "k-words", or simply words when k is clear). These oligonucleotides are hybridizing to an unknown DNA fragment, whose sequence one would like to determine.

Under ideal conditions, this target molecule will hybridize to all words whose Watson-Crick complements occur somewhere along its sequence. Thus, in principle, one would determine in a single microarray reaction the set of all k-long substrings of the target and try to infer the sequence from those data.

The average length of a uniquely resconstructible sequence using an 8-mer array is <200 bases, far below a single read length on commercial gel-lane machine. The main weakness of sequencing by hybridization is ambiguous solutions-when several sequences have the same spectrum; there is no way to determine the true sequence.

## Pal Nyren's Method

In 1996, Pal Nyren's group reported that natural nucleotide can be used to obtain efficient incorporation during a sequencing-by-synthesis protocol. The detection was based on the pyrophosphate (inorganic biphosphate) released during the DNA polymerase reaction, the quantitative conversion of pyrophosphate to ATP by sulfurylase and the subsequent production of visible light by firefly luciferase. The first major improvement was inclusion of dATPaS in place of dATP in the polymerization reaction, which enabled the pyro sequencing reaction to be performed in homogeneous phase in real time.

The non-specific signals were attributed to the fact that dATP is a substrate for luciferase. Conversely, dATPaS was found to be inert for luciferase, yet could be incorporated efficiently by all DNA polymerases tested. The second improvement was the introduction apyrase to the reaction to make a four-enzyme system. Apyrase allows nucleotides to be added sequentially without any intermediate washing step.

Pyrosequencing nonelectrophoretic real-time DNA sequencing method is based on sequencing by synthesis based on the pyrophosphate (inorganic biphosphate) released during the DNA polymerase reaction. In a cascade of enzymatic reaction, visible light is generated that is proportional to the number of incorporated nucleotides. The cascade starts with a nucleic acid polymerization reaction in which inorganic bip-hosphate (PPi) is released as a result of nucleotide incorporation by polymerase.

The released Pipe is subsequently converted to ATP by ATP sulfurylase, which provides the energy to luciferase to oxidize luciferin and generate light. The light so generated is captured by a CCD camera and recorded in the form of peaks known as pyrogram (compared with electropherograms in Sanger's method). Because the added nucleotide is known the sequence of template can be determined.

Standard pyrosequencing uses the Klenow fragment of E. coli DNA pol I, which is relatively slow polymerase. The ATP sulfurylase used in pyrosequencing is a recombinant version from the yeast and the luciferase is from the American firefly. The overall reaction from polymerization to light detection takes place within three to four seconds at real time.

One pmol of DNA in a pyrosequencing reaction yields $6 \times 10^{11}$ ATP molecules which in turn, generate more than $6 \times 10^9$ photons at a wavelength of 560 nm. This amount of light is easily detected by a photodiode, photomultiplier tube or a CCD camera. Pyrosequencing technology has been further improved into array-based massively parallel microfluidic sequencing platform.

## Automatic DNA Sequencer

A variant of the above dideoxy-method was developed, which allowed the production of automatic sequencers. In this new approach, different fluorescent dyes are tagged either to the oligonucleotide primer (dye primers) in each of the four reaction tubes (blue for A, red for C, etc), or to each of the four ddNTPs (dye terminators) used in a single reaction tube: when four tubes are used, they are pooled.

After the PCR reaction is over, the reaction mixture is subjected to separation of synthesized fragments through electrophoresis. Depending upon the electrophoretic system used, whether slab gel electrophoresis or capillary electrophoresis, following two types of automatic sequencing systems have been designed.

Automated DNA sequencer and details of sample loop.

## Slab Gel Sequencing Systems

These systems make use of ultrathin (75 µm) slab gels and involve running of atleast 96 lanes per gel. In these systems, automation in sample loading of sequencing gels has also been achieved, by using a plexiglass block having wells that are same distance apart as the comb teeth cut in a porous membrane that is used as a comb for drawing samples by capillary action.

Each well in plexiglass block is filed with a sample (PCR dideoxy-reaction mixture), so that when the porous membrane comb is lowered onto the sample wells in the pexiglass the samples are drawn up automatically into the comb teeth by capillary action.

Using this approach of employing porous combs, automated loading of up to 192, 384 or 480 samples per gel has been achieved. The porous comb with the samples is placed between the glass plates of the gel apparatus above the flat surface of the polymerized gel and the samples are driven from the comb into the gel by electrophoresis.

## Capillary Gel Electrophoresis

In these systems, slab gel electrophoresis is replaced by capillary gel electrophoresis to analyse DNA samples. In these systems, instead of scanning DNA as it migrates through 96 lanes each in a series of 96 capillary tubes, DNA fragments pass are scanned. In the original models of the above old slab gel machines, gels must be poured and reagents frequently reloaded, interrupting the sequencing.

In capillary gel sequencing systems, on the other hand, the robot moves the DNA samples and reagents through the tubes continuously, requiring attention only once a day. The system produces a steady flow of data, each signal representing one of the four DNA bases (adenine, cytosine, guanine and thymine).

## Examples of Sequencing

Beginning in the late 1990s, the scientific community witnessed a remarkable climax of accomplishments related to DNA sequencing. In addition to the historic sequencing of the human genome, sequences have now been generated for the genomes of several key model organisms, including the mouse (Mus musculus); the rat (Rattus norvegicus); two fruit flies (Drosophila melanogaster and D. pseudoobscura); two roundworms (Caenorhabditis elegans and C. briggsae);

yeast (Saccharomyces cerevisiae) and several other fungi; a malaria-carrying mosquito (Anopheles gambiae) along with a malaria-causing parasite (Plasmodium falciparum); two sea squirts (Ciona savignyi and C. intestinalis); a long list of microbes; and a couple of plants, including mustard weed (Arabidopsis thaliana) and rice (Oryza sativa).

Sequencing work is well underway on the honey bee (Apis mellifera), and is just getting started or expected to begin soon on the chimpanzee (Pan troglodytes), the cow (Bos taurus), the dog (Canis familiaris) and the chicken (Gallus gallus).

# Applications of DNA Sequencing

## Haplotype Phasing using High throughput Sequence Data

Resolution of haplotype phase is important to understanding shared disease-associate chromosomal segments containing variants that tend to be inherited en bloc, compound heterozygous (two or more risk alleles in one gene) and oligogenic (two or more risk alleles in multiple genes) genotype-phenotype associations, regulatory effects of genetic variation, and differential parent of origin effects in disease assocation studies. Furthermore, large databases of phased sequence data will be important resources for genome-wide association studies that utilize imputation, or estimation of genotypes not assayed by other technologies such as chip-based genotyping. This practice has become commonplace as investigators combine datasets to improve power to detect disease associations of small magnitude, and will be important for investigating rare variant effects. Short-read high throughput sequence data alone does not provide information about haplotype phase. However, several statistical algorithms based on pedigree information, common population haplotypes, and paired short reads have been developed that are applicable to high-throughput sequence data. Moreover, several investigators have developed experimental methods for haplotype phasing based on sorting individual metaphase chromosomes and subsequent sequencing, or from a combination of long-insert cloning and next-generation sequencing.

## High-throughput Sequencing and Mendelian Disease Genetics

The utility of high-throughput sequencing for investigation of disease genetics is great. The application of NGS for the identification of cardiovascular disease-associated loci has resulted in several notable successes, including the identification of BAG3 mutations as a cause of dilated cardiomyopathy, mutations in SMAD3 associated with familial aortic aneurysms, and AARS2 and ACAD9 in familial mitochondrial cardiomyopathy. These studies have provided intriguing hypotheses for follow-up work characterizing novel pathways in human cardiovascular disease. The genetic basis for several non-cardiovascular diseases has similarly been explored using exome and whole-genome sequencing. Notably, two studies have demonstrated the promise of NGS in aiding clinical diagnosis and management. Choi et al used exome sequencing to identify a mutation in SLC26A3 in a patient with the suspected renal salt-wasting Bartter syndrome; this finding allowed them to make the unanticipated diagnosis of congenital chloride diarrhea and modify clinical care accordingly. Worthey et al used exome sequencing to identify a missense mutation in the gene XIAP in a patient with intractable Crohn's-like inflammatory bowel disease, establishing a diagnosis of X-linked inhibitor of apoptosis (XIAP) deficiency. Subsequent allogeneic stem cell transplant resulted in dramatic improvement in the patient's gastrointestinal disease.

Thus far, these studies have focused on well-characterized diseases with extreme phenotypic manifestations and co-segregation analysis of single gene loci. However, filtering variants by co-segregation with the disease phenotype, as well as by comparison with population controls, e.g., the dbSNP and 1000 genomes genetic variation databases, has not always yielded a definitive answer. This difficulty is further compounded by the inclusion of non-validated SNPs in recent iterations of these databases. Consequently, these repositories now contain a small but definite subset of putative variants that are actually sequencing errors. Filtering of variants in co-segregation studies by their presence in these databases may thus lead to the misidentification of damaging mutations as benign polymorphisms. Further annotation of variants by identity-by-descent status, which seeks to identify common ancestral disease-associated haplotypes, represents an evolution of the co-segregation approach.

## High throughput Sequencing and Complex Disease Genetics

More recently there has been increasing interest in the use of high-throughput sequencing for association analysis with complex disease. Much of the focus has been on discovering the source of "missing heritability" of common complex diseases. Six years has elapsed since the first publication of a genome wide association study of common genetic variants and common disease. Since then, hundreds of highly statistically significant, replicated associations with common disease have been found. However, most alleles identified via this technique confer modest risk, and the heritability of common disease explained by these alleles in isolation or aggregate is low. One of the hypotheses for this relative paucity of high-effect associations is that rare variants of large effect contribute in aggregate to common disease. By virtue of their rarity, these variants have not been included on current genotyping arrays and therefore previous GWAS studies have thus been unable to assess their association with disease. Furthermore, several investigators have hypothesized that some of the modest associations between common variants and common disease are mediated via weak linkage disequilibrium between common marker variants and rare, causative variants of large effect. Recently, Stefansson, et al used a combination of large-scale chip-based genotyping and intermediate-depth (10 times) whole genome sequencing of a smaller cohort of cases and controls to identify a rare variant in a novel locus strongly associated with sick sinus syndrome. Importantly, this is the first demonstration of the use of whole genome sequencing to identify an association between a rare variant and complex disease. Though it is not yet cost effective to perform deep whole genome sequencing of large cohorts of individuals with common disease, as sequencing costs drop, genotype-phenotype association studies using whole genome sequencing may become feasible. Meanwhile, several efforts are currently underway to identify coding variants using exome sequencing associated with complex phenotypes.

An important advantage of whole genome over whole exome association studies is the ability to interrogate the noncoding genome. Despite systematic over-representation of protein coding regions on many genotyping arrays, 88% of significant GWAS associations are located in intronic or intergenic regions. Thus, exome-targeted sequencing approaches are likely to miss the majority of significant genome wide associations with common disease. For Mendelian disorders, the majority of underlying allelic variants identified thus far disrupt coding regions, and thus many early sequencing efforts have focused on the exome. However, it is likely that more comprehensive variant discovery will be required for discovery of many genotype-phenotype associations.

## Genome Sequencing and the Clinic

Applying the bulk of genetic predictive information to whole genome sequence data from individuals is one of the most difficult tasks in NGS data interpretation. We previously developed and applied a methodology for interpretation of genetic and environmental risk in a single subject using a combination of traditional clinical assessment, whole genome sequencing, and integration of genetic and environmental risk factors, and have recently done so for a family quartet. A similar approach has been applied to carrier testing for severe recessive childhood disease risk using NGS and for detection of fetal aneuploidy via sequencing of maternal blood samples. One of the main challenges to the widespread application of these analytical schemes is incomplete and inconsistent status of publicly-available genome annotation databases. Several annotation sources exist for gene regions, including the consensus coding sequence (CCDS) database, RefSeq, the UCSC Known Genes database, and the GENCODE and ENSEMBL databases. Each has advantages in terms of coverage and accuracy; however, the inconsistent use of these data in the literature is an issue for replicating research findings. Similarly, several variant databases exist for associations with Mendelian disorders, including the Human Gene Mutation Database, the Online Mendelian Inheritance in Man, and many disease-specific databases. None are well suited to variant-level annotation of whole genome sequence data and many contain annotation errors and common polymorphisms, by some estimates comprising approximately > 25% of the entries. Furthermore, these databases are contaminated by descriptions of susceptibility loci of questionable impact, and mutation annotations are often based on differing builds of the reference genome or outdated gene and protein sequences. Several prediction algorithms exist for predicting variant pathogenicity that are based on different combinations of evolutionary conservation, structural prediction, and physical properties of amino acid substitutions. However, they are limited in specificity and sensitivity, and concordance between predictions from the various algorithms is low. Databases for common variant – common disease associations and pharmacogenomic associations are more complete, but there is a great need for comprehensive, easily searchable, and accurate variant-level association databases as whole genome sequence data becomes more widely available.

## Applications of High throughput Sequencing

Though high-throughput sequencing has become synonymous with whole genome and exome sequencing, there are many other emerging applications for the technology. The first of these is whole transcriptome sequencing, which uses massively parallel sequencing to sequence RNA transcripts in various physiological conditions. This unique application allows for determination of allele-specific expression, information about alternative splicing, RNA editing events, and, via read depth, accurate quantification of messenger RNA (mRNA) copy number, and, therefore, gene expression. Compared with oligonucleotide expression arrays, RNAseq is able to quantify transcript abundance with a greater dynamic range and accuracy at extremes of transcript abundance, allowing for more accurate quantification of gene expression and rich functional genomics information. Matkovich et al, recently used a unique combination of RNAseq and a new technology, RNA-induced silencing complexes (RISC)-sequencing, to characterize cardiac mRNA regulation by microRNAs, small noncoding RNAs that regulate diverse cellular functions by facilitating mRNA degradation or inhibiting translation. Technologies that do not require generation and amplification of a cDNA library, such as the Helicos platform, are particularly well suited to this application because they require no prior knowledge of the transcriptome and avoid biases in gene expression measurements and sequencing errors that are related to reverse transcription.

Secondly, subsets of exomes can be queried in a high-throughput manner in the next generation of candidate gene studies using custom oligonucleotide based capture techniques coupled with high throughput sequencing. Combined with a pooled case-control approach, these study designs may prove to be valuable to gene finding or comprehensive sequence interrogation ("fine mapping") of genomic regions that have been linked with inherited disease by other technologies such as array-based genotyping or repetitive element mapping.

Third, there is increasing focus on the use of high-throughput sequencing in clinical diagnosis via the rapid identification of cell-free DNA. Specific to cardiovascular medicine, is the recent demonstration of the use of cell-free sequencing of blood samples for the identification of an organ-specific "transplant DNA" signature correlating with acute cellular rejection in a pilot study of heart transplant recipients. With confirmation in larger cohorts, technologies such as these may be combined with other functional assays of the genome such as gene expression arrays to obviate the need for endomyocardial biopsy surveillance in select patients.

There is an entirely separate dimension of heritable information that researchers are just beginning to explore on a genome-wide scale. Epigenetic traits, or heritable traits that do not involve DNA sequence changes, are often due to chemical modifications of the DNA molecule such as cytosine methylation in CpG regions. To date, bisulfite sequencing, in which 5-methyl-cytosine bases are converted to uracil by bisulfite and subsequently sequenced, identifying CpG regions with high uracil content that correspond to methyl-cytosine bases, has been the standard technique. However, single molecule sequencers yield polymerase kinetics information that correlates with methylation status and other structural information such as DNA polymerase footprint and RNA and DNA secondary structure.

## Uncovering the Secrets of our Past

Next generation sequencing allows what was previously technically impossible. DNA derived from human specimens thousands of years old can now be sequenced to unveil insights into our evolutionary past. One recent study gives insights into how the Americas may have become populated. Over 16,000 years ago, a land bridge connected Siberia to Alaska, allowing migration of the first people from Asia to the Americas. There are many theories as to how people populated the North and South of America. However, recent next generation sequencing data has suggested an interesting model. In an ancient burial site in Alaska, two preserved infants were found and their mitochondrial genome (the genetic sequence of the energy-producing organelle found in every cell in the body and is maternally inherited) was sequenced. Interestingly, the mitochondrial DNA from the two infants was very different from that of modern people living in North America—and was more closely similar to mitochondrial DNA from a 500 year-old Incan child mummy found in Argentina. This has huge implications as to how humans populated the Americas, suggesting that the people who first arrived in Alaska from Asia were already very diverse and that they traversed all the way down the American continent to South America.

## Checking the Validity of Herbal Supplements

Many people are turning to herbal supplements, seeking a more natural way to stay healthy. This has led to the dietary supplement market ballooning to an estimated worth of over $123 billion dollars worldwide. Although widely consumed, data from next generation sequencing has questioned whether some herbal supplements actually contained the ingredients on the label. Researchers at the University of Guelph in Ontario, Canada used DNA barcoding to sequence the contents of 44 herbal supplements. Their results were horrifying: a third of the samples did not even contain the advertised plant, with many containing ingredients not listed on the bottle. Some even contained other supplements that were instead powerful laxatives or were mixed with substances that can induce an allergic reaction in many people, such as nuts. Only two of the products tested contained 100% authentic ingredients. Luckily, largely due to the results of this sequencing study, legislators are starting to investigate the accuracy of the labeling of supplements in a bid to protect the public from products that, theoretically, should be improving their health.

According to the National Center for Complimentary and Integrative Health, 38% of all Americans use some form complementary medicine to help allay their ills. The use of alternative medicine receives a large amount of scepticism; however, researchers are using sequencing technologies to better understand how alternative medicine may influence our bodies. Ayurvedic medicine is based on a whole-body holistic approach. One form of Ayurveda, called Prakriti, focuses on the interplay between lifestyle and environment and how this may influence disease. Interestingly, an analysis of genetic markers (single nucleotide polymorphisms or SNPs) was able to accurately categorize a population of Indian males into the three groups used in Prakriti. For example, one group classified as Pitta, could correlate with the mutational profile of the gene, PGM1, which is at the center of many metabolic pathways. This suggests that the ancient practice of Ayurveda may have a genetic basis and scientifically validates this ancient medicine.

## Conserving Our Wildlife

Sequencing is also aiding conservation efforts around the world. For example, Thermo Fisher Scientific recently donated a Genetic Analyzer to the Cheetah Conservation Fund (CCF), Life Technologies Conservation Genetics Laboratory: the only fully equipped genetics lab at a conservation facility in Africa. Due to human-wildlife conflict, the cheetah population has declined by 90% in the 20th century alone, and the International Union for the Conservation of Nature (IUCN) lists them as "vulnerable." Sequencing enables the researchers at CCF to better understand the underlying genetics of this beautiful big cat and monitor the reproduction and health of the cheetah population. This information is crucial to ensure that we do not lose the Cheetah forever.

Every year, poachers kill 10% of the African elephant population, and many of the elephants are hunted for their ivory tusks. Researchers are using sequencing technology to track where elephants are being poached. By sequencing the genomes of tusks seized by law enforcement, they found that 27 out of 28 seizures occurred in one of four areas and most seizures were concentrated in just two regions. This work showed the location of the poaching "hotspots" and may allow enforcement officials to concentrate their efforts in those locations. Thus, sequencing technology is helping to protect our elephants from poachers and stem the illegal trade of ivory worldwide.

Even in the past three decades, DNA sequencing has gone from a very expensive and time-consuming process (the human genome project cost over \$3 billion and took over a decade to complete at the turn of the century) to a relatively inexpensive and fast technique (Next Generation Sequencing can sequence an entire genome for much less and in hours, not years). These advances have opened up new avenues for research that were previously thought impossible—like analyzing the genome of a single cell. Whether it is helping tackle human diseases, unveiling the secrets of our past, or bringing together ancient medicine with modern science, DNA sequencing has yielded some amazing discoveries, with even more in store in the coming years.

## Forensics

The ability to use low concentrations of DNA to obtain reliable sequencing reads has been extremely useful to the forensic scientist. In particular, the potential to sequence every DNA within a sample is attractive, especially since a crime scene often contains genetic material from multiple people. HTS is slowly being adopted in many forensics labs for human identification. In addition, recent advances allow forensic scientists to sequence the exome of a person after death, especially to determine the cause of death. For instance, death due to poisoning will show changes to the exome in affected organs. On the other hand, DNA sequencing can also determine that the deceased had a preexisting genetic ailment or predisposition. The challenges in this field include the development of extremely reliable analysis software, especially since the results of HTS cannot be manually examined.

## Genomics

Genome is the study of the structure, function, and inheritance of the genome (entire set of genetic material) of an organism. A major part of genomics is determining the sequence of molecules that make up the genomic deoxyribonucleic acid (DNA) content of an organism. The genomic DNA sequence is contained within an organism's chromosomes, one or more sets of which are found in each cell of an organism. The chromosomes can be further described as containing the fundamental units of heredity, the genes. Genes are transcriptional units, those regions of chromosomes that under appropriate circumstances are capable of producing a ribonucleic acid (RNA) transcript that can be translated into molecules of protein.

Every organism contains a basic set of chromosomes, unique in number and size for every species that includes the complete set of genes plus any DNA between them. While the term genome was not brought into use until 1920, the existence of genomes has been known since the late 19th century, when chromosomes were first observed as stained bodies visible under the microscope. The initial discovery of chromosomes was then followed in the 20th century by the mapping of genes on chromosomes based on the frequency of exchange of parts of chromosomes by a process called chromosomal crossing over, an event that occurs as a part of the normal process of recombination and the production of sex cells (gametes) during meiosis. The genes that could be mapped by chromosomal crossing over were mainly those for which mutant phenotypes (visible manifestations of an organism's genetic composition) had been observed, only a small proportion of the total genes in the genome. The discipline of genomics arose when the technology became available to deduce the complete nucleotide sequence of genomes, sequences generally in the range of billions of nucleotide pairs.

# Sequencing and Bioinformatic Analysis of Genomes

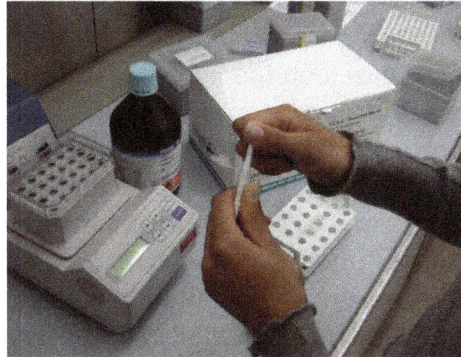

The process of DNA extraction is necessary to isolate molecules of DNA from cells or tissues.
A series of steps, including the use of protease enzymes to strip proteins from the DNA, are required
for isolating pure DNA that is suitable for use in later procedures, such as cloning or sequencing.

Genomic sequences are usually determined using automatic sequencing machines. In a typical experiment to determine a genomic sequence, genomic DNA first is extracted from a sample of cells of an organism and then is broken into many random fragments. These fragments are cloned in a DNA vector (carrier) that is capable of carrying large DNA inserts. Because the total amount of DNA that is required for sequencing and additional experimental analysis is several times the total amount of DNA in an organism's genome, each of the cloned fragments is amplified individually by replication inside a living bacterial cell, which reproduces rapidly and in great quantity to generate many bacterial clones. The cloned DNA is then extracted from the bacterial clones and is fed into the sequencing machine. The resulting sequence data are stored in a computer. When a large enough number of sequences from many different clones is obtained, the computer ties them together using sequence overlaps. The result is the genomic sequence, which is then deposited in a publicly accessible database.

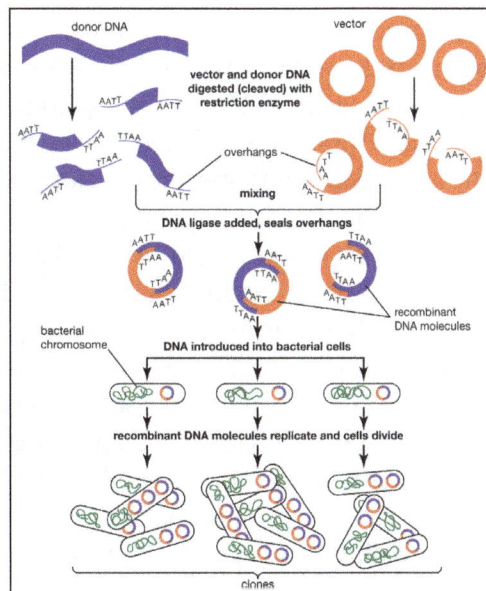

In genomics research, fragments of genomic DNA are inserted into a vector and amplified by
replication in bacterial cells. In this way, large amounts of DNA can be cloned and extracted from
the bacterial cells. The DNA is then sequenced and further analyzed using bioinformatics techniques.

A complete genomic sequence in itself is of limited use; the data must be processed to find the genes and, if possible, their associated regulatory sequences. The need for these detailed analyses has given rise to the field of bioinformatics, in which computer programs scan DNA sequences looking for genes, using algorithms based on the known features of genes, such as unique triplet sequences of nucleotides known as start and stop codons that span a gene-sized segment of DNA or sequences of DNA that are known to be important in regulating adjacent genes. Once candidate genes are identified, they must be annotated to ascribe potential functions. Such annotation is generally based on known functions of similar gene sequences in other organisms, a type of analysis made possible by evolutionary conservation of gene sequence and function across organisms as a result of their common ancestry. However, after annotation there is still a subset of genes for which functions cannot be deduced; these functions gradually become revealed with further research.

## Genomics Applications

## Functional Genomics

Analysis of genes at the functional level is one of the main uses of genomics, an area known generally as functional genomics. Determining the function of individual genes can be done in several ways. Classical, or forward, genetic methodology starts with a randomly obtained mutant of interesting phenotype and uses this to find the normal gene sequence and its function. Reverse genetics starts with the normal gene sequence (as obtained by genomics), induces a targeted mutation into the gene, then, by observing how the mutation changes phenotype, deduces the normal function of the gene. The two approaches, forward and reverse, are complementary. Often a gene identified by forward genetics has been mapped to one specific chromosomal region, and the full genomic sequence reveals a gene in this position with an already annotated function.

## Gene Identification by Microarray Genomic Analysis

Genomics has greatly simplified the process of finding the complete subset of genes that is relevant to some specific temporal or developmental event of an organism. For example, microarray technology allows a sample of the DNA of a clone of each gene in a whole genome to be laid out in order on the surface of a special chip, which is basically a small thin piece of glass that is treated in such a way that DNA molecules firmly stick to the surface. For any specific developmental stage of interest (e.g., the growth of root hairs in a plant or the production of a limb bud in an animal), the total RNA is extracted from cells of the organism, labeled with a fluorescent dye, and used to bathe the surfaces of the microarrays. As a result of specific base pairing, the RNAs present bind to the genes from which they were originally transcribed and produce fluorescent spots on the chip's surface. Hence, the total set of genes that were transcribed during the biological function of interest can be determined. Note that forward genetics can aim at a similar goal of assembling the subset of genes that pertain to some specific biological process. The forward genetic approach is to first induce a large set of mutations with phenotypes that appear to change the process in question, followed by attempts to define the genes that normally guide the process. However, the technique can only identify genes for which mutations produce an easily recognizable mutant phenotype, and so genes with subtle effects are often missed.

## Comparative Genomics

A further application of genomics is in the study of evolutionary relationships. Using classical genetics, evolutionary relationships can be studied by comparing the chromosome size, number, and banding patterns between populations, species, and genera. However, if full genomic sequences are available, comparative genomics brings to bear a resolving power that is much greater than that of classical genetics methods and allows much more subtle differences to be detected. This is because comparative genomics allows the DNAs of organisms to be compared directly and on a small scale. Overall, comparative genomics has shown high levels of similarity between closely related animals, such as humans and chimpanzees, and, more surprisingly, similarity between seemingly distantly related animals, such as humans and insects. Comparative genomics applied to distinct populations of humans has shown that the human species is a genetic continuum, and the differences between populations are restricted to a very small subset of genes that affect superficial appearance such as skin color. DNA sequence can be measured mathematically genomic analysis can be quantified in a very precise way to measure specific degrees of relatedness. Genomics has detected small-scale changes, such as the existence of surprisingly high levels of gene duplication and mobile elements within genomes.

## References

- DNAEngID30001ES: atlasgeneticsoncology.org, Retrieved 14 June, 2019

- Sequencing, science: whatisbiotechnology.org, Retrieved 24 April, 2019

- DNA-Sequencing-Fact-Sheet, about-genomics: genome.gov, Retrieved 21 January, 2019

- PMC3364518: ncbi.nlm.nih.gov, Retrieved 7 August, 2019

- Dna-sequencing-7-methods-used-for-dna-sequencing, dna-sequencing, dna: biologydiscussion.com, Retrieved 3 April, 2019

- Decoding-the-genome-applications-of-dna-sequencing: bitesizebio.com, Retrieved 13 February, 2019

- Dna-sequencing: biologydictionary.net, Retrieved 9 May, 2019

- Genomics, science: britannica.com, Retrieved 31 July, 2019

# Chapter 5

## Gene Expression, Regulation and Translation

The process through which information from a gene gets used to synthesize a functional gene product is known as gene expression. The mechanisms which act to repress or induce genes are called gene regulation. This chapter discusses in detail the concepts and processes related to the expression, regulation and translation of genes in eukaryotes and prokaryotes.

## Gene Expression

Genes encode proteins and proteins dictate cell function. Therefore, the thousands of genes expressed in a particular cell determine what that cell can do. Moreover, each step in the flow of information from DNA to RNA to protein provides the cell with a potential control point for self-regulating its functions by adjusting the amount and type of proteins it manufactures.

At any given time, the amount of a particular protein in a cell reflects the balance between that protein's synthetic and degradative biochemical pathways. On the synthetic side of this balance, recall that protein production starts at transcription (DNA to RNA) and continues with translation (RNA to protein). Thus, control of these processes plays a critical role in determining what proteins are present in a cell and in what amounts. In addition, the way in which a cell processes its RNA transcripts and newly made proteins also greatly influences protein levels.

### Regulation of Gene Expression

The amounts and types of mRNA molecules in a cell reflect the function of that cell. In fact, thousands of transcripts are produced every second in every cell. Given this statistic, it is not surprising that the primary control point for gene expression is usually at the very beginning of the protein production process — the initiation of transcription. RNA transcription makes an efficient control point because many proteins can be made from a single mRNA molecule.

Transcript processing provides an additional level of regulation for eukaryotes, and the presence of a nucleus makes this possible. In prokaryotes, translation of a transcript begins before the transcript is complete, due to the proximity of ribosomes to the new mRNA molecules. In eukaryotes, however, transcripts are modified in the nucleus before they are exported to the cytoplasm for translation.

Eukaryotic transcripts are also more complex than prokaryotic transcripts. For instance, the primary transcripts synthesized by RNA polymerase contain sequences that will not be part of the mature RNA. These intervening sequences are called introns, and they are removed before the mature mRNA leaves the nucleus. The remaining regions of the transcript, which include the protein-coding regions, are called exons, and they are spliced together to produce the mature mRNA. Eukaryotic transcripts are also modified at their ends, which affects their stability and translation.

Of course, there are many cases in which cells must respond quickly to changing environmental conditions. In these situations, the regulatory control point may come well after transcription. For example, early development in most animals relies on translational control because very little transcription occurs during the first few cell divisions after fertilization. Eggs therefore contain many maternally originated mRNA transcripts as a ready reserve for translation after fertilization.

On the degradative side of the balance, cells can rapidly adjust their protein levels through the enzymatic breakdown of RNA transcripts and existing protein molecules. Both of these actions result in decreased amounts of certain proteins. Often, this breakdown is linked to specific events in the cell. The eukaryotic cell cycle provides a good example of how protein breakdown is linked to cellular events. This cycle is divided into several phases, each of which is characterized by distinct cyclin proteins that act as key regulators for that phase. Before a cell can progress from one phase of the cell cycle to the next, it must degrade the cyclin that characterizes that particular phase of the cycle. Failure to degrade a cyclin stops the cycle from continuing.

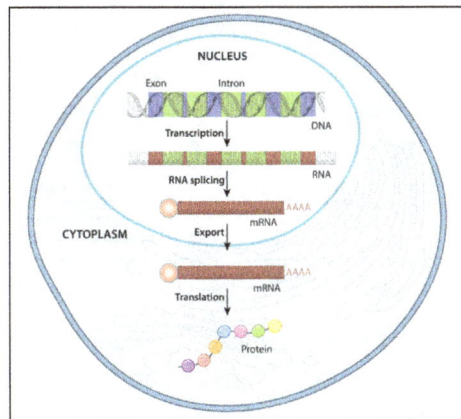

Overview of the flow of information from DNA to protein in a eukaryote.

First, both coding and noncoding regions of DNA are transcribed into mRNA. Some regions are removed (introns) during initial mRNA processing. The remaining exons are then spliced together, and the spliced mRNA molecule (red) is prepared for export out of the nucleus through addition of an endcap (sphere) and a polyA tail. Once in the cytoplasm, the mRNA can be used to construct a protein.

## Expression of the Genes Needed by Cells

Only a fraction of the genes in a cell are expressed at any one time. The variety of gene expression profiles characteristic of different cell types arise because these cells have distinct sets of transcription regulators. Some of these regulators work to increase transcription, whereas others prevent or suppress it.

Normally, transcription begins when an RNA polymerase binds to a so-called promoter sequence on the DNA molecule. This sequence is almost always located just upstream from the starting point for transcription (the 5' end of the DNA), though it can be located downstream of the mRNA (3' end). In recent years, researchers have discovered that other DNA sequences, knownas enhancer sequences, also play an important part in transcription by providing binding sites for regulatory proteins that affect RNA polymerase activity. Binding of regulatory proteins to an enhancer sequence causes a shift in chromatin structure that either promotes or inhibits RNA polymerase and transcription factor binding. A more open chromatin structure is associated with active gene transcription. In contrast, a more compact chromatin structure is associated with transcriptional inactivity.

Some regulatory proteins affect the transcription of multiple genes. This occurs because multiple copies of the regulatory protein binding sites exist within the genome of a cell. Consequently, regulatory proteins can have different roles for different genes, and this is one mechanism by which cells can coordinate the regulation of many genes at once.

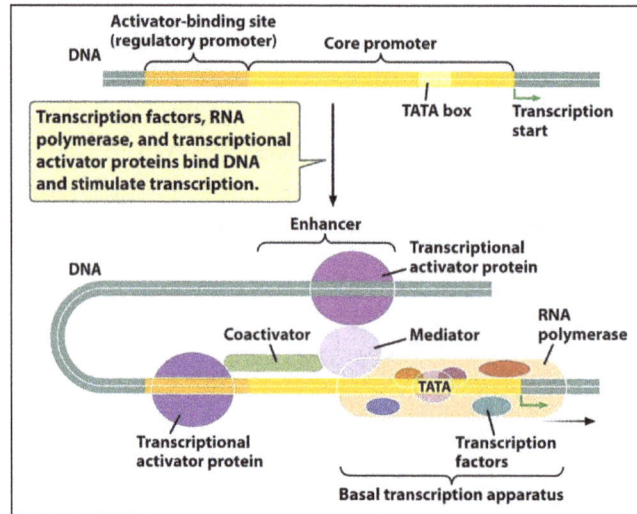

Modulation of transcription.

An activator protein bound to DNA at an upstream enhancer sequence can attract proteins to the promoter region that activate RNA polymerase (green) and thus transcription. The DNA can loop around on itself to cause this interaction between an activator protein and other proteins that mediate the activity of RNA polymerase.

## Increase and Decrease in Gene Expression in Response to Environmental Change

In prokaryotes, regulatory proteins are often controlled by nutrient availability. This allows organisms such as bacteria to rapidly adjust their transcription patterns in response to environmental conditions. In addition, regulatory sites on prokaryotic DNA are typically located close to transcription promoter sites — and this plays an important part in gene expression.

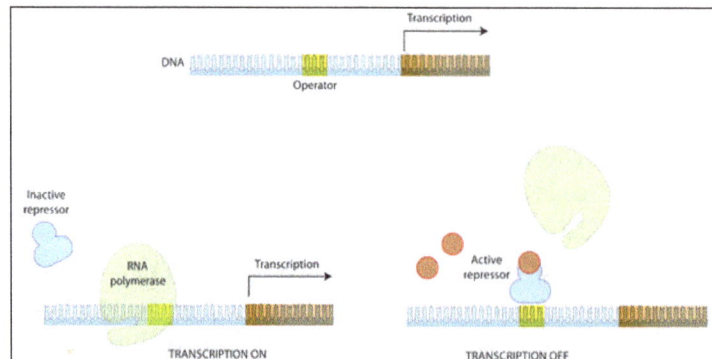

Transcription repression near the promoter region.

Molecules can interfere with RNA polymerase binding. An inactive repressor protein (blue) can become activated by another molecule (red circle). This active repressor can bind to a region near the promoter called an operator (yellow) and thus interfere with RNA polymerase binding to the promoter, effectively preventing transcription.

For an example of how this works, imagine a bacterium with a surplus of amino acids that signal the turning "on" of some genes and the turning "off" of others. In this particular example, cells might want to turn "on" genes for proteins that metabolize amino acids and turn "off" genes for proteins that synthesize amino acids. Some of these amino acids would bind to positive regulatory proteins called activators. Activator proteins bind to regulatory sites on DNA nearby to promoter regions that act as on/off switches. This binding facilitates RNA polymerase activity and transcription of nearby genes. At the same time, however, other amino acids would bind to negative regulatory proteins called repressors, which in turn bind to regulatory sites in the DNA that effectively block RNA polymerase binding.

The control of gene expression in eukaryotes is more complex than that in prokaryotes. In general, a greater number of regulatory proteins are involved, and regulatory binding sites may be located quite far from transcription promoter sites. Also, eukaryotic gene expression is usually regulated by a combination of several regulatory proteins acting together, which allows for greater flexibility in the control of gene expression.

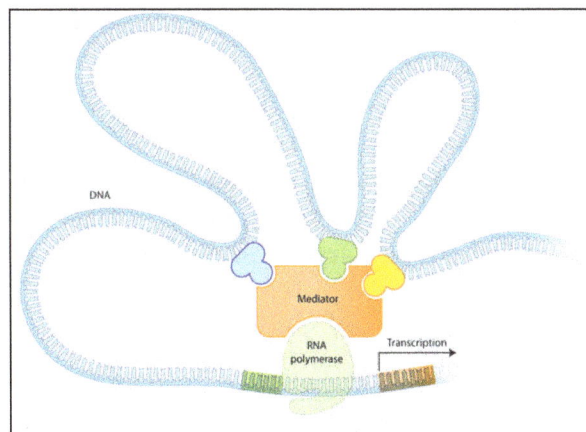

The complexity of multiple regulators

Transcriptional regulators can each have a different role. Combinations of one, two, or three regulators (blue, green, and yellow shapes) can affect transcription in different ways by differentially affecting a mediator complex (orange), which is also composed of proteins. The effect is that the same gene can be transcribed in multiple ways, depending on the combination, presence, or absence of various transcriptional regulator proteins.

The enhancer sequences are DNA sequences that are bound by an activator protein, and they can be located thousands of base pairs away from a promoter, either upstream or downstream from a gene. Activator protein binding is thought to cause DNA to loop out, bringing the activator protein into physical proximity with RNA polymerase and the other proteins in the complex that promote the initiation of transcription.

Different cell types express characteristic sets of transcriptional regulators. In fact, as multicellular organisms develop, different sets of cells within these organisms turn specific combinations of regulators on and off. Such developmental patterns are responsible for the variety of cell types present in the mature organism.

The wide variety of cell types in a single organism can depend on different transcription factor activity in each cell type. Different transcription factors can turn on at different times during successive generations of cells. As cells mature and go through different stages (arrows), transcription factors (colored balls) can act on gene expression and change the cell in different ways. This change affects the next generation of cells derived from that cell. In subsequent generations, it is the combination of different transcription factors that can ultimately determine cell type.

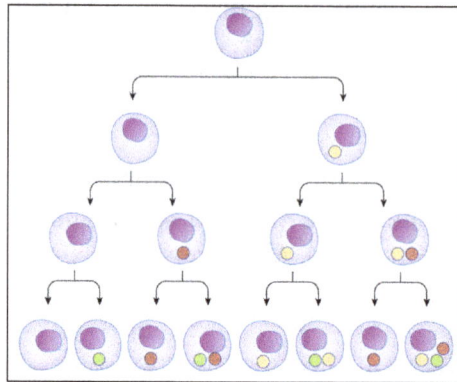

Transcriptional regulators can determine cell types.

To live, cells must be able to respond to changes in their environment. Regulation of the two main steps of protein production — transcription and translation — is critical to this adaptability. Cells can control which genes get transcribed and which transcripts get translated; further, they can biochemically process transcripts and proteins in order to affect their activity. Regulation of transcription and translation occurs in both prokaryotes and eukaryotes, but it is far more complex in eukaryotes.

# Gene Expression Analysis

Figure: Siamese cats have colored points because of a temperature-sensitive pigmentation gene. In cooler areas of a cat's body (nose and paws), this gene is expressed to a greater degree.

Gene expression is dynamic, and the same gene may act in different ways under different circumstances. For example, imagine that two organisms have similar genotypes but different phenotypes. What is the cause of this variation in phenotype? Could the difference stem from differing regulation of gene expression? Could temperature affect expression of the organism's DNA? Might some other factor be responsible? To go about answering these types of questions, researchers often use laboratory techniques such as a Northern blot or serial analysis of gene expression (SAGE). Both of these techniques make it possible to identify which genes are turned on and which are turned off within cells. Subsequently, this information can be used to help determine what circumstances trigger expression of various genes.

Both Northern blots and SAGE analyses work by measuring levels of mRNA, the intermediary between DNA and protein. Remember, in order to activate a gene, a cell must first copy the DNA sequence of that gene into a piece of mRNA known as a transcript. Thus, by determining which mRNA transcripts are present in a cell, scientists can determine which genes are expressed in that cell at different stages of development and under different environmental conditions.

## Northern Blots

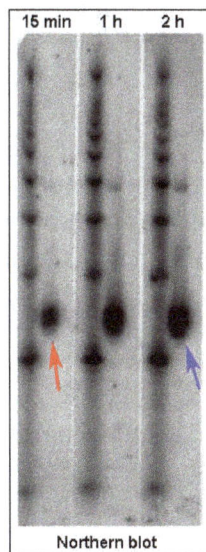

Northern blot

A Northern blot allows comparison of mRNA levels, which are a direct reflection of gene expression. This blot above shows the tracking of mRNA levels in a cell sample at various intervals after exposure to a drug. At 15 minutes after exposure, mRNA levels are low (red arrow, small black mark); at two hours after exposure, these levels have increased (blue arrow, larger black mark).

The quantity of mRNA transcript for a single gene directly reflects how much transcription of that gene has occurred. Tracking of that quantity will therefore indicate how vigorously a gene is transcribed, or expressed. To visualize differences in the quantity of mRNA produced by different groups of cells or at different times, researchers often use the method known as a Northern blot. For this method, researchers must first isolate mRNA from a biological sample by exposing the cells within it to a protease, which is an enzyme that breaks down cell membranes and releases the genetic material in the cells. Next, the mRNA is separated from the DNA, proteins, lipids, and other cellular contents. The different fragments of mRNA are then separated from one another via gel electrophoresis (a technique that separates molecules by passing an electrical current through a gel medium containing the molecules) and transferred to a filter or other solid support using a technique known as blotting. To identify the mRNA transcripts produced by a particular gene, the researchers next incubate the sample with a short piece of single-stranded RNA or DNA (also known as a probe) that is labeled with a radioactive molecule. Designed to be complementary to mRNA from the gene of interest, the probe will bind to this sequence. Later, when the filter is placed against X-ray film, the radioactivity in the probe will expose the film, thereby making marks on it. The intensity of the resulting marks, called bands, tells researchers how much mRNA was in the sample, which is a direct indicator of how strongly the gene of interest is expressed.

## Study the Expression of Multiple Genes Simultaneously

Until recently, scientists studied gene expression by looking at only one or very few gene transcripts at a time. Thankfully, new techniques now make large-scale studies of gene expression possible. One such technique is SAGE (serial analysis of gene expression). A method for measuring the expression patterns of many genes at once, SAGE not only allows scientists to analyze thousands of gene transcripts simultaneously, but it also enables them to determine which genes are active in different tissues or at different stages of cellular development.

## Working of SAGE

SAGE identifies and counts the mRNA transcripts in a cell with the help of short snippets of the genetic code, called tags. These tags, which are a maximum of 14 nucleotides long, enable researchers to match an mRNA transcript with the specific gene that produced it. In most cases, each tag contains enough information to uniquely identify a transcript. The name "serial analysis" refers to the fact that tags are read sequentially as a continuous string of information.

## Capturing mRNA

To begin a SAGE analysis, researchers must first separate the mRNA in a sample from the other cellular contents. To do this, they attach long strips of thymine nucleotides to tiny magnetic beads. When researchers flush the contents of a cell over the beads, these thymine strips form complementary base pairs with the poly-A tails of the mRNA molecules. Thus, when the flushing process is complete, the mRNA transcripts from the sample are captured because they are attached to the magnetic beads, while the other contents of the cells flush past the beads and are discarded.

## Rewriting mRNA into cDNA

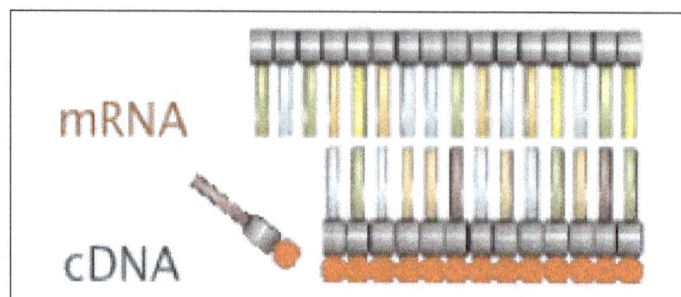

Figure: Reverse transcription converts mRNA into cDNA.

mRNA is more fragile than DNA, which makes it difficult to handle and analyze. To solve this problem, researchers often convert mRNA samples into complementary DNA sequences, or cDNA. This is done by reversing the natural process a cell uses to make mRNA from DNA, a method known as reverse transcription. The reverse transcription process doesn't use DNA polymerase or RNA polymerase; instead, it employs a special enzyme called reverse transcriptase. This enzyme makes cDNA sequences that are complementary to each mRNA transcript, essentially creating a converted form of the same sequence. This new single-stranded cDNA is then converted into a double-stranded cDNA molecule.

## Cutting Tags from each cDNA

To begin the next portion of SAGE, the researchers use a cutting enzyme to slice off short segments of nucleotides, called tags, at designated positions in each cDNA molecule. Next, two tags from each cDNA are combined into a single unit. These tags then become the representative for the gene they came from, and they act as a unique identifier in the form of a stand-in. Without having to process the entire mRNA sequence thereafter, scientists can use these shorter tag sequences to keep track of whether a specific gene was expressed in mRNA form.

## Linking Tags together in Chains for Sequencing

After the different tags have been made from each mRNA sequence, they are next linked together into long chains called concatemers. These concatemers therefore contain representatives of mRNAs from a group of genes. Linking the tags together in a concatemer is important, because it means that researchers will later be able to read thousands of tags at once during the analysis portion of the SAGE procedure.

## Copying and Reading the Chains

Although the researchers now possess concatemers representing the genes expressed in the sample, they need multiple copies of these concatemers if they wish to run the molecules through a sequencing machine. Thus, just before sequencing, the concatemers are inserted into bacteria, and through their own replication process these bacteria make millions of copies of each concatemer chain. This step increases the volume of material, and it therefore ensures that there is a baseline amount of material necessary for a sequencing machine. After that, researchers use a sequencing machine to decode and read the long string of nucleotides in each chain.

## Identifying and Counting the Tags

Finally, a computer processes the data from the sequencing machine and compiles a list of tags. By comparing the tags to a sequence database, the researchers can identify the mRNA (and ultimately the gene) that each tag came from. By subsequently counting the number of times each tag is observed, the researchers can also estimate the degree to which a particular gene is expressed: the more often a tag appears, the greater the level of gene expression.

## Researchers Learn from SAGE

Compared to other techniques for measuring gene expression, SAGE offers a significant advantage because it measures the expression of both known and unknown genes. Sometimes, when analyzing SAGE data, computers cannot find matches for certain tags in their sequence databases. What does this mean? Interestingly, a lack of matches indicates that the mRNA used to produce these tags is associated with genes that have not been studied before. In this way, SAGE has been used to discover new genes involved in a variety of diseases.

## Other ways to Measure Gene Expression

In addition to Northern blot tests and SAGE analyses, there are several other techniques for analyzing gene expression. Most of these techniques, including microarray analysis and reverse transcription polymerase chain reaction (RT-PCR), work by measuring mRNA levels. However, researchers can also analyze gene expression by directly measuring protein levels with a technique known as a Western blot.

# Prokaryotic Gene Regulation

The DNA of prokaryotes is organized into a circular chromosome supercoiled in the nucleoid region of the cell cytoplasm. Proteins that are needed for a specific function are encoded together in blocks called operons. For example, all of the genes needed to use lactose as an energy source are coded next to each other in the lactose (or lac) operon.

In prokaryotic cells, there are three types of regulatory molecules that can affect the expression of operons: repressors, activators, and inducers. Repressors are proteins that suppress transcription of a gene in response to an external stimulus, whereas activators are proteins that increase the transcription of a gene in response to an external stimulus. Finally, inducers are small molecules that either activate or repress transcription depending on the needs of the cell and the availability of substrate.

## Gene Regulation in Prokaryotes

In bacteria and archaea, structural proteins with related functions—such as the genes that encode the enzymes that catalyze the many steps in a single biochemical pathway—are usually encoded together within the genome in a block called an operon and are transcribed together under the control of a single promoter. This forms a polycistronic transcript. The promoter then has simultaneous control over the regulation of the transcription of these structural genes because they will either all be needed at the same time, or none will be needed.

In prokaryotes, structural genes of related function are often organized together on the genome and transcribed together under the control of a single promoter. The operon's regulatory region includes both the promoter and the operator. If a repressor binds to the operator, then the structural genes will not be transcribed. Alternatively, activators may bind to the regulatory region, enhancing transcription.

French scientists François Jacob (1920–2013) and Jacques Monod at the Pasteur Institute were the first to show the organization of bacterial genes into operons, through their studies on the lac operon of E. coli. They found that in E. coli, all of the structural genes that encode enzymes needed to use lactose as an energy source lie next to each other in the lactose (or lac) operon under the control of a single promoter, the lac promoter. For this work, they won the Nobel Prize in Physiology or Medicine in 1965.

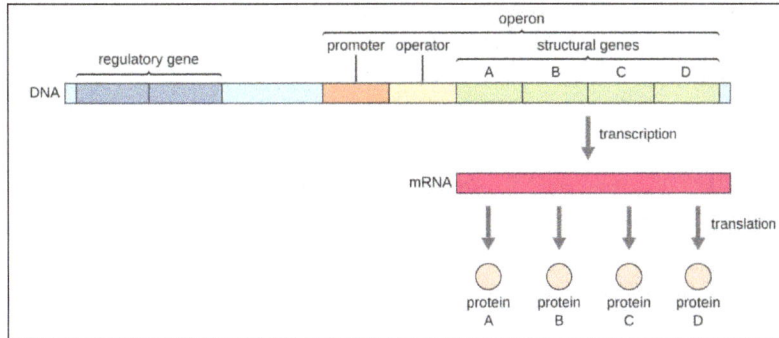

Although eukaryotic genes are not organized into operons, prokaryotic operons are excellent models for learning about gene regulation generally. There are some gene clusters in eukaryotes that function similar to operons. Many of the principles can be applied to eukaryotic systems and contribute to our understanding of changes in gene expression in eukaryotes that can result pathological changes such as cancer.

Each operon includes DNA sequences that influence its own transcription; these are located in a region called the regulatory region. The regulatory region includes the promoter and the region surrounding the promoter, to which transcription factors, proteins encoded by regulatory genes, can bind. Transcription factors influence the binding of RNA polymerase to the promoter and allow its progression to transcribe structural genes. A repressor is a transcription factor that suppresses transcription of a gene in response to an external stimulus by binding to a DNA sequence within the regulatory region called the operator, which is located between the RNA polymerase binding site of the promoter and the transcriptional start site of the first structural gene. Repressor binding physically blocks RNA polymerase from transcribing structural genes. Conversely, an activator is a transcription factor that increases the transcription of a gene in response to an external stimulus by facilitating RNA polymerase binding to the promoter. An inducer, a third type of regulatory molecule, is a small molecule that either activates or represses transcription by interacting with a repressor or an activator.

In prokaryotes, there are examples of operons whose gene products are required rather consistently and whose expression, therefore, is unregulated. Such operons are constitutively expressed, meaning they are transcribed and translated continuously to provide the cell with constant intermediate levels of the protein products. Such genes encode enzymes involved in housekeeping functions required for cellular maintenance, including DNA replication, repair, and expression, as well as enzymes involved in core metabolism. In contrast, there are other prokaryotic operons that are expressed only when needed and are regulated by repressors, activators, and inducers.

## Trp Operon: A Repressor Operon

Bacteria such as E. coli need amino acids to survive. Tryptophan is one such amino acid that E.

coli can ingest from the environment. E. coli can also synthesize tryptophan using enzymes that are encoded by five genes. These five genes are next to each other in what is called the tryptophan (trp) operon. If tryptophan is present in the environment, then E. coli does not need to synthesize it and the switch controlling the activation of the genes in the trp operon is switched off. However, when tryptophan availability is low, the switch controlling the operon is turned on, transcription is initiated, the genes are expressed, and tryptophan is synthesized.

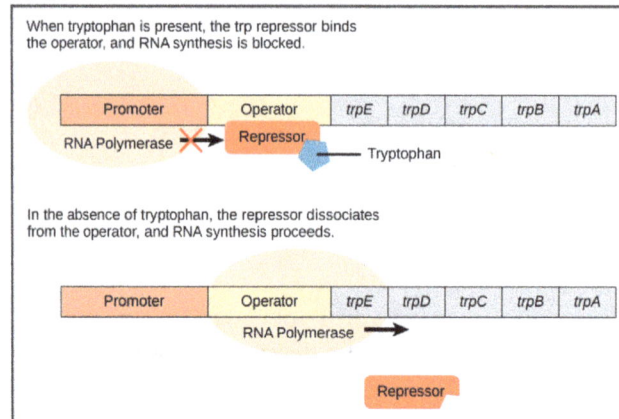

The five genes that are needed to synthesize tryptophan in E. coli are located next to each other in the trp operon. When tryptophan is plentiful, two tryptophan molecules bind the repressor protein at the operator sequence. This physically blocks the RNA polymerase from transcribing the tryptophan genes. When tryptophan is absent, the repressor protein does not bind to the operator and the genes are transcribed.

A DNA sequence that codes for proteins is referred to as the coding region. The five coding regions for the tryptophan biosynthesis enzymes are arranged sequentially on the chromosome in the operon. Just before the coding region is the transcriptional start site. This is the region of DNA to which RNA polymerase binds to initiate transcription. The promoter sequence is upstream of the transcriptional start site; each operon has a sequence within or near the promoter to which proteins (activators or repressors) can bind and regulate transcription.

A DNA sequence called the operator sequence is encoded between the promoter region and the first trp coding gene. This operator contains the DNA code to which the repressor protein can bind. When tryptophan is present in the cell, two tryptophan molecules bind to the trp repressor, which changes shape to bind to the trp operator. Binding of the tryptophan–repressor complex at the operator physically prevents the RNA polymerase from binding, and transcribing the downstream genes.

When tryptophan is not present in the cell, the repressor by itself does not bind to the operator; therefore, the operon is active and tryptophan is synthesized. Because the repressor protein actively binds to the operator to keep the genes turned off, the trp operon is negatively regulated and the proteins that bind to the operator to silence trp expression are negative regulators.

## Catabolite Activator Protein: Activator Regulator

Just as the trp operon is negatively regulated by tryptophan molecules, there are proteins that bind to the operator sequences that act as a positive regulator to turn genes on and activate them.

For example, when glucose is scarce, E. coli bacteria can turn to other sugar sources for fuel. To do this, new genes to process these alternate sugars must be transcribed. When glucose levels drop, cyclic AMP (cAMP) begins to accumulate in the cell. The cAMP molecule is a signaling molecule that is involved in glucose and energy metabolism in E. coli. When glucose levels decline in the cell, accumulating cAMP binds to the positive regulator catabolite activator protein (CAP), a protein that binds to the promoters of operons that control the processing of alternative sugars. When cAMP binds to CAP, the complex binds to the promoter region of the genes that are needed to use the alternate sugar sources. In these operons, a CAP binding site is located upstream of the RNA polymerase binding site in the promoter. This increases the binding ability of RNA polymerase to the promoter region and the transcription of the genes.

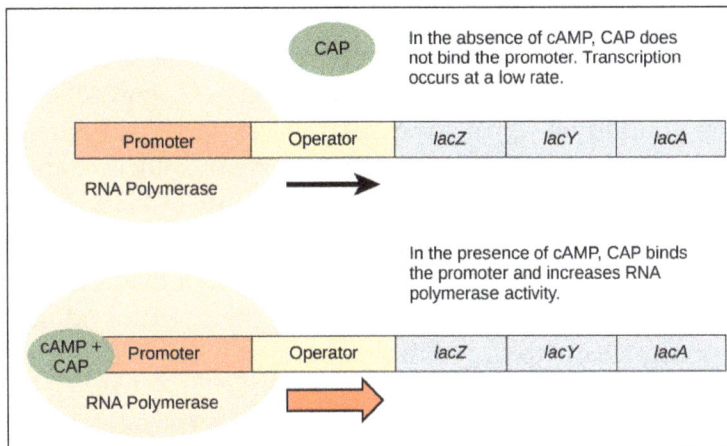

When glucose levels fall, E. coli may use other sugars for fuel but must transcribe new genes to do so. As glucose supplies become limited, cAMP levels increase. This cAMP binds to the CAP protein, a positive regulator that binds to an operator region upstream of the genes required to use other sugar sources.

## Lac Operon: Inducer Operon

The third type of gene regulation in prokaryotic cells occurs through inducible operons, which have proteins that bind to activate or repress transcription depending on the local environment and the needs of the cell. The lac operon is a typical inducible operon. As mentioned previously, E. coli is able to use other sugars as energy sources when glucose concentrations are low. To do so, the cAMP–CAP protein complex serves as a positive regulator to induce transcription. One such sugar source is lactose. The lac operon encodes the genes necessary to acquire and process the lactose from the local environment. CAP binds to the operator sequence upstream of the promoter that initiates transcription of the lac operon. However, for the lac operon to be activated, two conditions must be met. First, the level of glucose must be very low or non-existent. Second, lactose must be present. Only when glucose is absent and lactose is present will the lac operon be transcribed. This makes sense for the cell, because it would be energetically wasteful to create the proteins to process lactose if glucose was plentiful or lactose was not available.

If glucose is absent, then CAP can bind to the operator sequence to activate transcription. If lactose is absent, then the repressor binds to the operator to prevent transcription. If either of these requirements is met, then transcription remains off. Only when both conditions are satisfied is the lac operon transcribed.

| Table: Signals that Induce or Repress Transcription of the lac Operon | | | | |
|---|---|---|---|---|
| Glucose | CAP binds | Lactose | Repressor binds | Transcription |
| + | – | – | + | No |
| + | – | + | – | Some |
| – | + | – | + | No |
| – | + | + | – | Yes |

# Eukaryotic Gene Regulation

Although the control of gene expression is far more complex in eukaryotes than in bacteria, the same basic principles apply. The expression of eukaryotic genes is controlled primarily at the level of initiation of transcription, although in some cases transcription may be attenuated and regulated at subsequent steps. As in bacteria, transcription in eukaryotic cells is controlled by proteins that bind to specific regulatory sequences and modulate the activity of RNA polymerase. The intricate task of regulating gene expression in the many differentiated cell types of multicellular organisms is accomplished primarily by the combined actions of multiple different transcriptional regulatory proteins. In addition, the packaging of DNA into chromatin and its modification by methylation impart further levels of complexity to the control of eukaryotic gene expression.

## Cis-acting Regulatory Sequences: Promoters and Enhancers

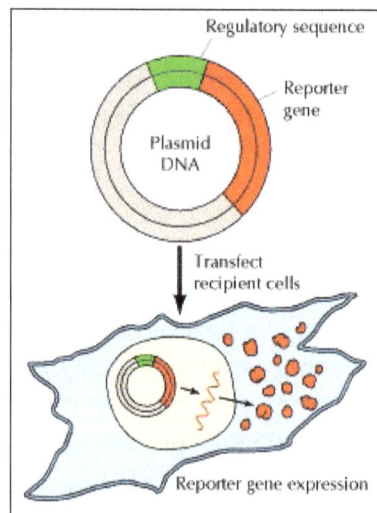

Figure: Identification of eukaryotic regulatory sequences.

As discussed, transcription in bacteria is regulated by the binding of proteins to cis-acting sequences (e.g., the lac operator) that control the transcription of adjacent genes. Similar cis-acting sequences regulate the expression of eukaryotic genes. These sequences have been identified in mammalian cells largely by the use of gene transfer assays to study the activity of suspected regulatory regions of cloned genes. The eukaryotic regulatory sequences are usually ligated to a reporter gene that encodes an easily detectable enzyme. The expression of the reporter gene following its transfer into cultured cells then provides a sensitive assay for the ability of the cloned regulatory sequences to direct transcription. Biologically active regulatory regions can thus be identified, and in vitro mutagenesis can be used to determine the roles of specific sequences within the region.

The regulatory sequence of a cloned eukaryotic gene is ligated to a reporter gene that encodes an easily detectable enzyme. The resulting plasmid is then introduced into cultured recipient cells by transfection. An active regulatory sequence directs transcription of the reporter gene, expression of which is then detected in the transfected cells.

Genes transcribed by RNA polymerase II have two core promoter elements, the TATA box and the Inr sequence, that serve as specific binding sites for general transcription factors. Other cis-acting sequences serve as binding sites for a wide variety of regulatory factors that control the expression of individual genes. These cis-acting regulatory sequences are frequently, though not always, located upstream of the TATA box. For example, two regulatory sequences that are found in many eukaryotic genes were identified by studies of the promoter of the herpes simplex virus gene that encodes thymidine kinase. Both of these sequences are located within 100 base pairs upstream of the TATA box: Their consensus sequences are CCAAT and GGGCGG (called a GC box). Specific proteins that bind to these sequences and stimulate transcription have since been identified.

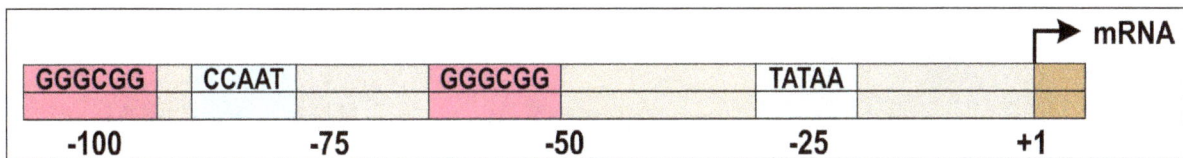

Figure: Eukaryotic promoter.

The promoter of the thymidine kinase gene of herpes simplex virus contains three sequence elements upstream of the TATA box that are required for efficient transcription: a CCAAT box and two GC boxes (consensus sequence GGGCGG).

In contrast to the relatively simple organization of CCAAT and GC boxes in the herpes thymidine kinase promoter, many genes in mammalian cells are controlled by regulatory sequences located farther away (sometimes more than 10 kilobases) from the transcription start site. These sequences, called enhancers, were first identified by Walter Schaffner in 1981 during studies of the promoter of another virus, SV40 (Figure). In addition to a TATA box and a set of six GC boxes, two 72-base-pair repeats located farther upstream are required for efficient transcription from this promoter. These sequences were found to stimulate transcription from other promoters as well as from that of SV40, and, surprisingly, their activity depended on neither their distance nor their orientation with respect to the transcription initiation site. They could stimulate transcription when placed either upstream or downstream of the promoter, in either a forward or backward orientation.

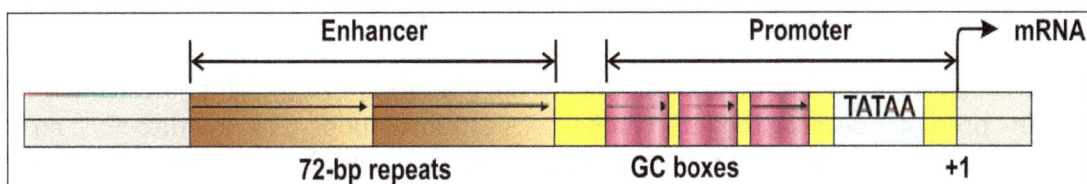

Figure: The SV40 enhancer.

The SV40 promoter for early gene expression contains a TATA box and six GC boxes arranged in three sets of repeated sequences. In addition, efficient transcription requires an upstream enhancer consisting of two 72-base-pair (bp) repeats.

Without an enhancer, the gene is transcribed at a low basal level (A). Addition of an enhancer, E—for example, the SV40 72-base-pair repeats—stimulates transcription. The enhancer is active not only when placed just upstream of the promoter (B), but also when inserted up to several kilobases either upstream or downstream from the transcription start site (C and D). In addition, enhancers are active in either the forward or backward orientation (E).

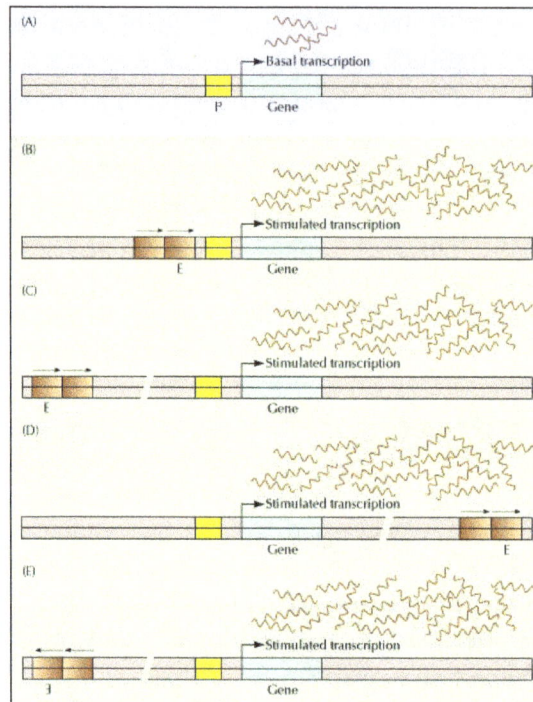

Figure: Action of enhancers.

The ability of enhancers to function even when separated by long distances from transcription initiation sites at first suggested that they work by mechanisms different from those of promoters. However, this has turned out not to be the case: Enhancers, like promoters, function by binding transcription factors that then regulate RNA polymerase. This is possible because of DNA looping, which allows a transcription factor bound to a distant enhancer to interact with RNA polymerase or general transcription factors at the promoter. Transcription factors bound to distant enhancers can thus work by the same mechanisms as those bound adjacent to promoters, so there is no fundamental difference between the actions of enhancers and those of cis-acting regulatory sequences adjacent to transcription start sites. Interestingly, although enhancers were first identified in mammalian cells, they have subsequently been found in bacteria—an unusual instance in which studies of eukaryotes served as a model for the simpler prokaryotic systems.

Transcription factors bound at distant enhancers are able to interact with general transcription factors at the promoter because the intervening DNA can form loops. There is therefore no fundamental difference between the action of transcription factors bound to DNA just upstream of the promoter and to distant enhancers.

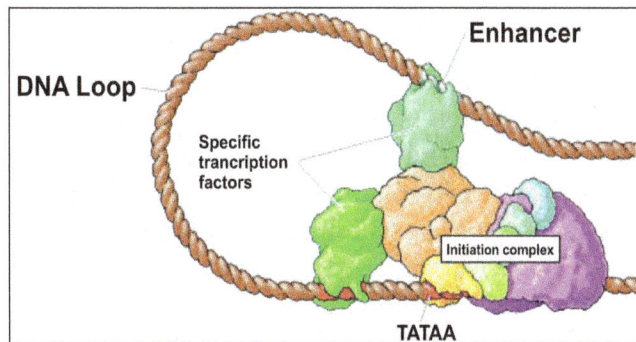
Figure: DNA looping

The binding of specific transcriptional regulatory proteins to enhancers is responsible for the control of gene expression during development and differentiation, as well as during the response of cells to hormones and growth factors. One of the most thoroughly studied mammalian enhancers controls the transcription of immunoglobulin genes in B lymphocytes. Gene transfer experiments have established that the immunoglobulin enhancer is active in lymphocytes, but not in other types of cells. Thus, this regulatory sequence is at least partly responsible for tissue-specific expression of the immunoglobulin genes in the appropriate differentiated cell type.

An important aspect of enhancers is that they usually contain multiple functional sequence elements that bind different transcriptional regulatory proteins. These proteins work together to regulate gene expression. The immunoglobulin heavy-chain enhancer, for example, spans approximately 200 base pairs and contains at least nine distinct sequence elements that serve as protein-binding sites. Mutation of any one of these sequences reduces but does not abolish enhancer activity, indicating that the functions of individual proteins that bind to the enhancer are at least partly redundant. Many of the individual sequence elements of the immunoglobulin enhancer by themselves stimulate transcription in nonlymphoid cells. The restricted activity of the intact enhancer in B lymphocytes therefore does not result from the tissue-specific function of each of its components. Instead, tissue-specific expression results from the combination of the individual sequence elements that make up the complete enhancer. These elements include some cis-acting regulatory sequences that bind transcriptional activators that are expressed specifically in B lymphocytes, as well as other regulatory sequences that bind repressors in nonlymphoid cells. Thus, the immunoglobulin enhancer contains negative regulatory elements that inhibit transcription in inappropriate cell types, as well as positive regulatory elements that activate transcription in B lymphocytes. The overall activity of the enhancer is greater than the sum of its parts, reflecting the combined action of the proteins associated with each of its individual sequence elements.

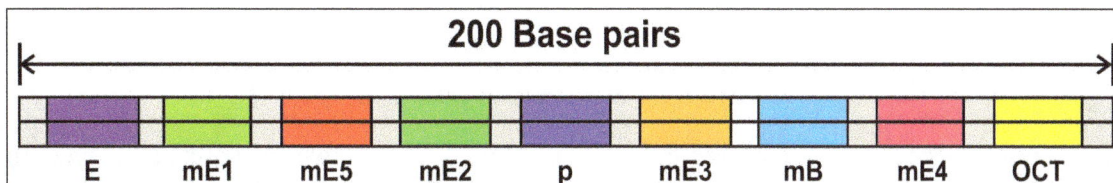
Figure: The immunoglobulin enhancer.

The immunoglobulin heavy-chain enhancer spans about 200 bases and contains nine functional sequence elements (E, µE1-5, π, µB, and OCT), which together stimulate transcription in B lymphocytes.

## Transcriptional Regulatory Proteins

The isolation of a variety of transcriptional regulatory proteins has been based on their specific binding to promoter or enhancer sequences. Protein binding to these DNA sequences is commonly analyzed by two types of experiments. The first, footprinting, was described earlier in connection with the binding of RNA polymerase to prokaryotic promoters. The second approach is the electrophoretic-mobility shift assay, in which a radiolabeled DNA fragment is incubated with a protein preparation and then subjected to electrophoresis through a nondenaturing gel. Protein binding is detected as a decrease in the electrophoretic mobility of the DNA fragment, since its migration through the gel is slowed by the bound protein. The combined use of footprinting and electrophoretic-mobility shift assays has led to the correlation of protein-binding sites with the regulatory elements of enhancers and promoters, indicating that these sequences generally constitute the recognition sites of specific DNA-binding proteins.

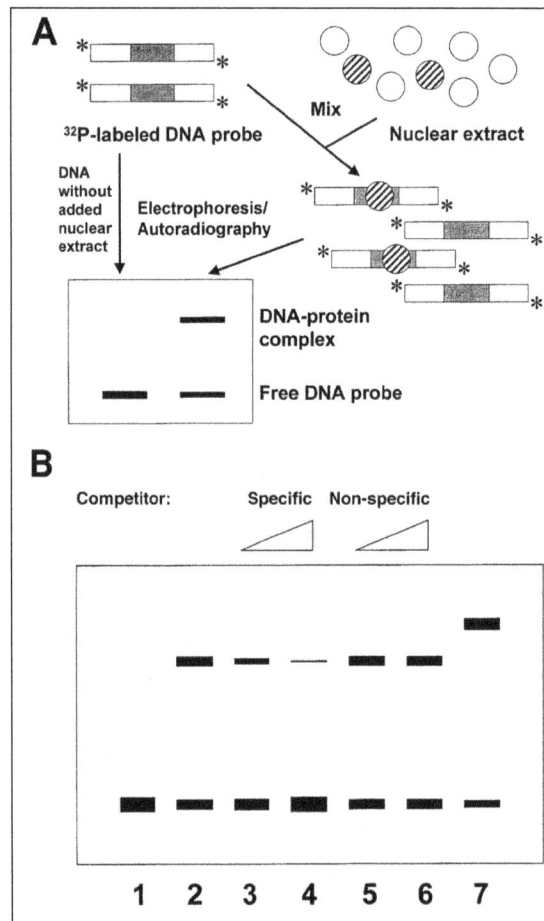

Figure: Electrophoretic-mobility shift assay.

A sample containing radiolabeled fragments of DNA is divided into two, and one half of the sample is incubated with a protein that binds to a specific DNA sequence. Samples are then analyzed by electrophoresis in a nondenaturing gel so that the protein remains bound to DNA. Protein binding is detected by the slower migration of DNA-protein complexes compared to that of free DNA. Only a fraction of the DNA in the sample is actually bound to protein, so both DNAprotein complexes and free DNA is detected following incubation of the DNA with protein.

One of the prototypes of eukaryotic transcription factors was initially identified by Robert Tjian and his colleagues during studies of the transcription of SV40 DNA. This factor (called Sp1, for specificity protein 1) was found to stimulate transcription from the SV40 promoter, but not from several other promoters, in cell-free extracts. Then, stimulation of transcription by Sp1 was found to depend on the presence of the GC boxes in the SV40 promoter: if these sequences were deleted, stimulation by Sp1 was abolished. Moreover, footprinting experiments established that Sp1 binds specifically to the GC box sequences. Taken together, these results indicate that the GC box represents a specific binding site for a transcriptional activator—Sp1. Similar experiments have established that many other transcriptional regulatory sequences, including the CCAAT sequence and the various sequence elements of the immunoglobulin enhancer, also represent recognition sites for sequence-specific DNA-binding proteins.

Table: Examples of Transcription Factors and Their DNA-Binding Sites.

| Transcription factor | Consensus binding site |
| --- | --- |
| Specificity protein 1 (Sp1) | GGGCGG |
| CCAAT/Enhancer binding protein (C/EBP) | CCAAT |
| Activator protein 1 (AP1) | TGACTCA |
| Octamer binding proteins | ATGCAAAT |
| (OCT-1 and OCT-2) | |
| E-box binding proteins (E12, E47, E2-2) | CANNTG[a] |

The specific binding of Sp1 to the GC box not only established the action of Sp1 as a sequence-specific transcription factor; it also suggested a general approach to the purification of transcription factors. The isolation of these proteins initially presented a formidable challenge because they are present in very small quantities (e.g., only 0.001% of total cell protein) that are difficult to purify by conventional biochemical techniques. This problem was overcome in the purification of Sp1 by DNA-affinity chromatography. Multiple copies of oligonucleotides corresponding to the GC box sequence were bound to a solid support, and cell extracts were passed through the oligonucleotide column. Because Sp1 bound to the GC box with high affinity, it was specifically retained on the column while other proteins were not. Highly purified Sp1 could thus be obtained and used for further studies, including partial determination of its amino acid sequence, which in turn led to cloning of the gene for Sp1.

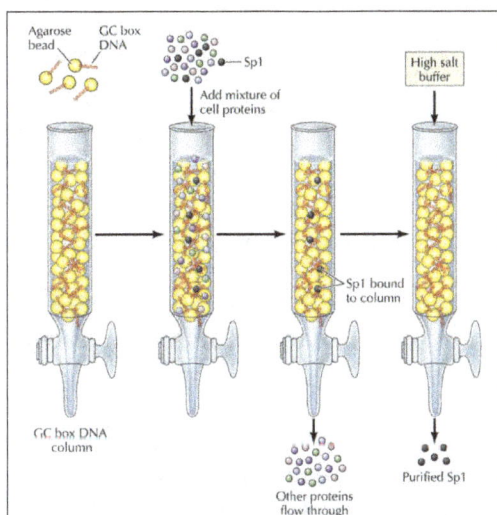

Figure: Purification of Sp1 by DNA-affinity chromatography.

A double-stranded oligonucleotide containing repeated GC box sequences is bound to agarose beads, which are poured into a column. A mixture of cell proteins containing Sp1 is then applied to the column; because Sp1 specifically binds to the GC box oligonucleotide, it is retained on the column while other proteins flow through. Washing the column with high salt buffer then dissociates Sp1 from the GC box DNA, yielding purified Sp1.

The general approach of DNA-affinity chromatography, first optimized for the purification of Sp1, has been used successfully to isolate a wide variety of sequence-specific DNA-binding proteins from eukaryotic cells. Protein purification has been followed by gene cloning and nucleotide sequencing, leading to the accumulation of a great deal of information on the structure and function of these critical regulatory proteins.

## Structure and Function of Transcriptional Activators

Because transcription factors are central to the regulation of gene expression, understanding the mechanisms of their action is a major area of ongoing research in cell and molecular biology. The most thoroughly studied of these proteins are transcriptional activators, which, like Sp1, bind to regulatory DNA sequences and stimulate transcription. In general, these factors have been found to consist of two domains: one region of the protein specifically binds DNA; the other activates transcription by interacting with other components of the transcriptional machinery. Transcriptional activators appear to be modular proteins, in the sense that the DNA binding and activation domains of different factors can frequently be interchanged using recombinant DNA techniques. Such manipulations result in hybrid transcription factors, which activate transcription by binding to promoter or enhancer sequences determined by the specificity of their DNA-binding domains. It therefore appears that the basic function of the DNA-binding domain is to anchor the transcription factor to the proper site on DNA; the activation domain then independently stimulates transcription by interacting with other proteins.

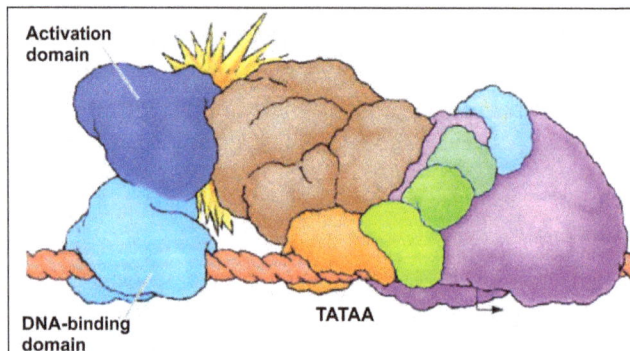

Figure: Structure of transcriptional activators.

Transcriptional activators consist of two independent domains. The DNA-binding domain recognizes a specific DNA sequence, and the activation domain interacts with other components of the transcriptional machinery.

Many different transcription factors have now been identified in eukaryotic cells, as might be expected, given the intricacies of tissue-specific and inducible gene expression in complex multicellular organisms. Molecular characterization has revealed that the DNA-binding domains of many of these proteins are related to one another. Zinc finger domains contain repeats of cysteine and histidine residues that bind zinc ions and fold into looped structures ("fingers") that bind DNA. These domains were initially identified in the polymerase III transcription factor TFIIIA but are also common among transcription factors that regulate polymerase II promoters, including Sp1. Other examples of transcription factors that contain zinc finger domains are the steroid hormone receptors, which regulate gene transcription in response to hormones such as estrogen and testosterone.

(A) Zinc finger domains consist of loops in which an α helix and a β sheet coordinately bind a zinc ion. (B) Helix-turn-helix domains consist of three (or in some cases four) helical regions. One helix (helix 3) makes most of the contacts with DNA, while helices 1 and 2 lie on top and stabilize the interaction. (C) The DNA-binding domains of leucine zipper proteins are formed from two distinct polypeptide chains. Interactions between the hydrophobic side chains of leucine residues exposed on one side of a helical region (the leucine zipper) are responsible for dimerization. Immediately following the leucine zipper is a DNA-binding helix, which is rich in basic amino acids. (D) Helix-loop-helix domains are similar to leucine zippers, except that the dimerization domains of these proteins each consist of two helical regions separated by a loop.

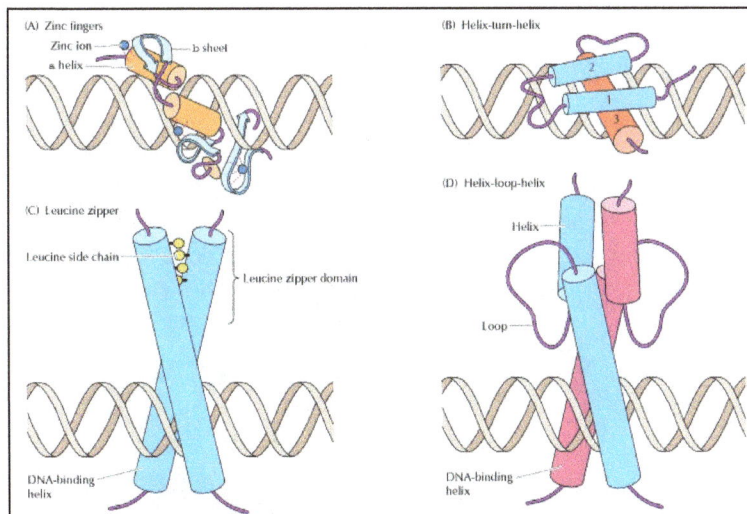

Figure: Families of DNA-binding domains.

The helix-turn-helix motif was first recognized in prokaryotic DNA-binding proteins, including the E. coli catabolite activator protein (CAP). In these proteins, one helix makes most of the contacts with DNA, while the other helices lie across the complex to stabilize the interaction. In eukaryotic cells, helix-turn-helix proteins include the homeodomain proteins, which play critical roles in the regulation of gene expression during embryonic development. The genes encoding these proteins were first discovered as developmental mutants in Drosophila. Some of the earliest recognized Drosophila mutants (termed homeotic mutants in 1894) resulted in the development of flies in which one body part was transformed into another. For example, in the homeotic mutant called Antennapedia, legs rather than antennae grow out of the head of the fly. Genetic analysis of these mutants, pioneered by Ed Lewis in the 1940s, has shown that Drosophila contains nine homeotic genes, each of which specifies the identity of a different body segment. Molecular cloning and analysis of these genes then indicated that they contain conserved sequences of 180 base pairs (called homeoboxes) that encode the DNA-binding domains (homeodomains) of transcription factors. A wide variety of additional homeodomain proteins have since been identified in fungi, plants, and other animals, including humans. Vertebrate homeobox genes are strikingly similar to their Drosophila counterparts in both structure and function, demonstrating the highly conserved roles of these transcription factors in animal development.

Two other families of DNA-binding proteins, leucine zipper and helix-loop-helix proteins, contain DNA-binding domains formed by dimerization of two polypeptide chains. The leucine zipper contains four or five leucine residues spaced at intervals of seven amino acids, resulting in their hydrophobic side chains being exposed at one side of a helical region. This region serves as the dimerization domain for the two protein subunits, which are held together by hydrophobic interactions between the leucine side chains. Immediately following the leucine zipper is a region rich in positively charged amino acids (lysine and arginine) that binds DNA. The helix-loop-helix proteins are similar in structure, except that their dimerization domains are each formed by two helical regions separated by a loop. An important feature of both leucine zipper and helix-loop-helix transcription factors is that different members of these families can dimerize with each other. Thus, the combination of distinct protein subunits can form an expanded array of factors that can differ both in DNA sequence recognition and in transcription-stimulating activities. Both leucine zipper and helix-loop-helix proteins play important roles in regulating tissue-specific and inducible gene expression, and the formation of dimers between different members of these families is a critical aspect of the control of their function.

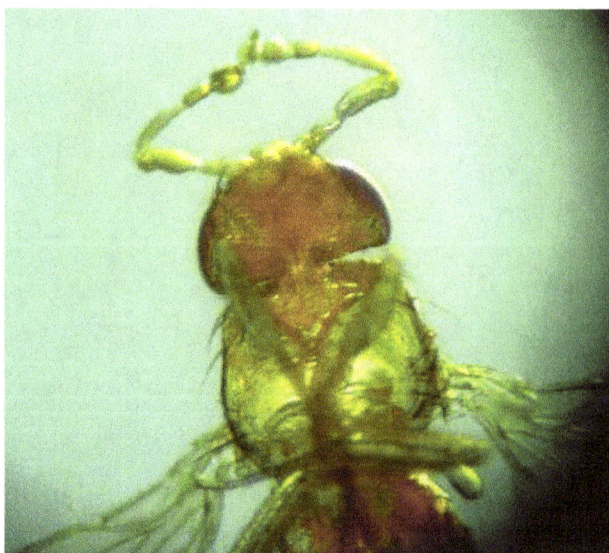

Figure: The Antennapedia mutant flies have legs growing out of their heads in place of antennae.

The activation domains of transcription factors are not as well characterized as their DNA-binding domains. Some, called acidic activation domains, are rich in negatively charged residues (aspartate and glutamate); others are rich in proline or glutamine residues. These activation domains are thought to stimulate transcription by interacting with general transcription factors, such as TFIIB or TFIID, thereby facilitating the assembly of a transcription complex on the promoter. For example, the activation domains of several transcription factors (including Sp1) have been shown to interact with TFIID by binding to TBP-associated factors (TAFs). An important feature of these interactions is that different activators can bind to different general transcription factors or TAFs, providing a mechanism by which the combined action of multiple factors can synergistically stimulate transcription—a key feature of transcriptional regulation in eukaryotic cells.

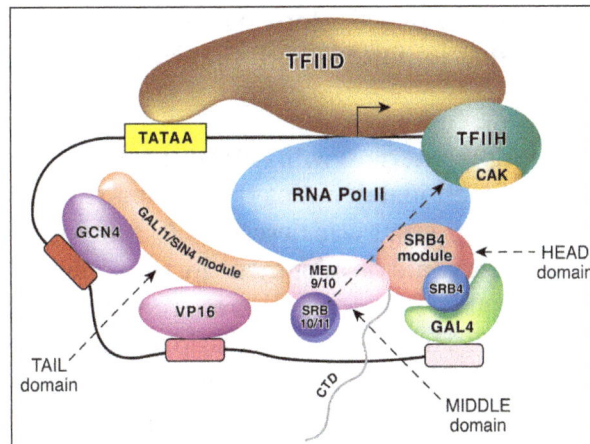

Figure: Synergistic action of transcriptional activators.

Different transcriptional activators can interact with the general transcription factor TFIID by binding to different TAFs.

## Eukaryotic Repressors

Gene expression in eukaryotic cells is regulated by repressors as well as by transcriptional activators. Like their prokaryotic counterparts, eukaryotic repressors bind to specific DNA sequences and inhibit transcription. In some cases, eukaryotic repressors simply interfere with the binding of other transcription factors to DNA. For example, the binding of a repressor near the transcription start site can block the interaction of RNA polymerase or general transcription factors with the promoter, which is similar to the action of repressors in bacteria. Other repressors compete with activators for binding to specific regulatory sequences. Some such repressors contain the same DNA-binding domain as the activator but lack its activation domain. As a result, their binding to a promoter or enhancer blocks the binding of the activator, thereby inhibiting transcription.

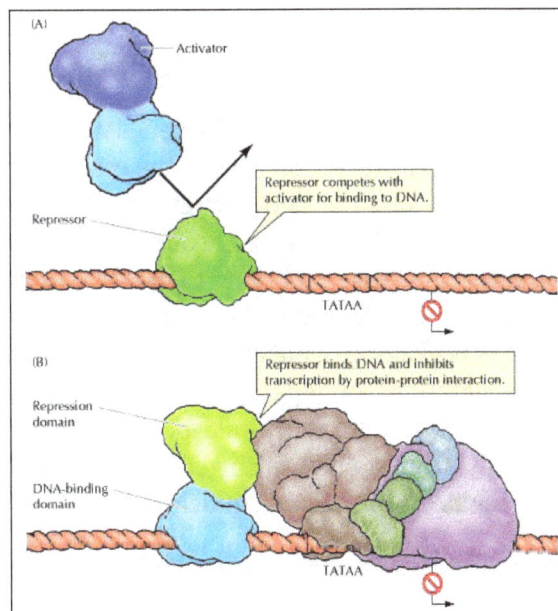

Figure: Action of eukaryotic repressors.

(A) Some repressors block the binding of activators to regulatory sequences. (B) Other repressors have active repression domains that inhibit transcription by interactions with general transcription factors.

In contrast to repressors that simply interfere with activator binding, many repressors (called active repressors) contain specific functional domains that inhibit transcription via protein-protein interactions. The first such active repressor was described in 1990 during studies of a gene called Kruppel, which is involved in embryonic development in Drosophila. Molecular analysis of the Kruppel protein demonstrated that it contains a discrete repression domain, which is linked to a zinc finger DNA-binding domain. The Krüppel repression domain could be interchanged with distinct DNA-binding domains of other transcription factors. These hybrid molecules also repressed transcription, indicating that the Krüppel repression domain inhibits transcription via protein-protein interactions, irrespective of its site of binding to DNA.

Many active repressors have since been found to play key roles in the regulation of transcription in animal cells, in many cases serving as critical regulators of cell growth and differentiation. As with transcriptional activators, several distinct types of repression domains have been identified. For example, the repression domain of Kruppel is rich in alanine residues, whereas other repression domains are rich in proline or acidic residues. The functional targets of repressors are also diverse. Some repressors inhibit transcription by interacting with general transcription factors, such as TFIID; others are thought to interact with specific activator proteins.

The regulation of transcription by repressors as well as by activators considerably extends the range of mechanisms that control the expression of eukaryotic genes. One important role of repressors may be to inhibit the expression of tissue-specific genes in inappropriate cell types. For example, a repressor-binding site in the immunoglobulin enhancer is thought to contribute to its tissue-specific expression by suppressing transcription in no lymphoid cell types. Other repressors play key roles in the control of cell proliferation and differentiation in response to hormones and growth factors.

## Relationship of Chromatin Structure to Transcription

In the preceding discussion, the transcription of eukaryotic genes was considered as if they were present within the nucleus as naked DNA. However, this is not the case. The DNA of all eukaryotic cells is tightly bound to histones, forming chromatin. The basic structural unit of chromatin is the nucleosome, which consists of 146 base pairs of DNA wrapped around two molecules each of histones H2A, H2B, H3, and H4, with one molecule of histone H1 bound to the DNA as it enters the nucleosome core particle. The chromatin is then further condensed by being coiled into higher-order structures organized into large loops of DNA. This packaging of eukaryotic DNA in chromatin clearly has important consequences in terms of its availability as a template for transcription, so chromatin structure is a critical aspect of gene expression in eukaryotic cells. Indeed, both activators and repressors regulate transcription in eukaryotes not only by interacting with general transcription factors and other components of the transcriptional machinery, but also by inducing changes in the structure of chromatin.

The relationship between chromatin structure and transcription is evident at several levels. First, actively transcribed genes are found in decondensed chromatin, corresponding to the extended 10 nm chromatin fibers. For example, microscopic visualization of the polytene chromosomes of

Drosophila indicates that regions of the genome that are actively engaged in RNA synthesis correspond to decondensed chromosome regions. Similarly, actively transcribed genes in vertebrate cells are present in a decondensed fraction of chromatin that is more accessible to transcription factors than is the rest of the genome.

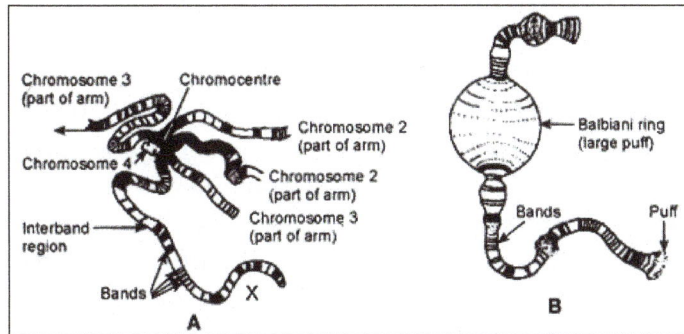

Figure: Decondensed chromosome regions in Drosophila.

Decondensation of chromatin, however, is not sufficient to make the DNA an accessible template for transcription. Even in decondensed chromatin, actively transcribed genes remain bound to histones and packaged in nucleosomes, so transcription factors and RNA polymerase are still faced with the problem of interacting with chromatin rather than with naked DNA. The tight winding of DNA around the nucleosome core particle is a major obstacle to transcription, affecting both the ability of transcription factors to bind DNA and the ability of RNA polymerase to transcribe through a chromatin template. This inhibitory effect of nucleosomes is relieved by acetylation of histones and by the binding of two nonhistone chromosomal proteins (called HMG-14 and HMG-17) to nucleosomes of actively transcribed genes. (HMG stands for high-mobility group proteins; these proteins migrate rapidly during gel electrophoresis). Additional proteins called nucleosome remodeling factors facilitate the binding of transcription factors to chromatin by altering nucleosome structure.

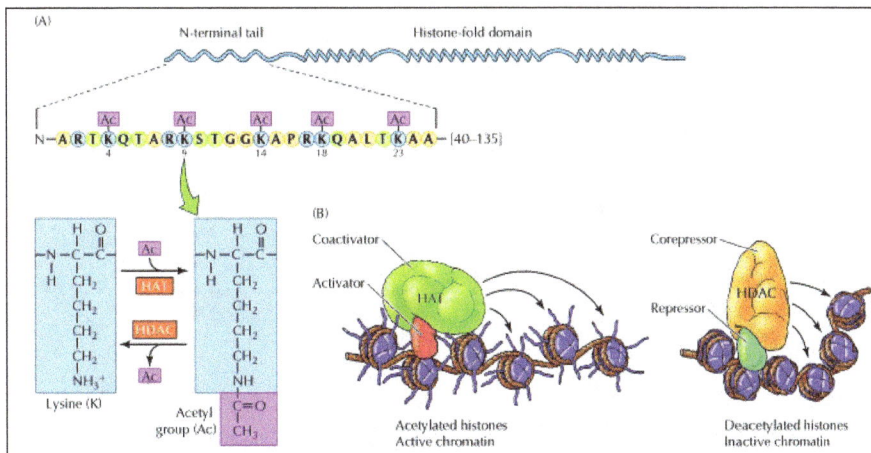

Figure: Histone acetylation.

Acetylation of histones has been correlated with transcriptionally active chromatin in a wide variety of cell types. The core histones (H2A, H2B, H3 and H4) have two domains: a histone fold domain, which is involved in interactions with other histones and in wrapping DNA around the nucleosome core particle, and an amino-terminal tail domain, which extends outside of the nucleosome. The amino-terminal tail domains are rich in lysine and can be modified by acetylation at specific lysine residues. Acetylation reduces the net positive charge of the histones, and may weaken their binding to DNA as well as altering their interactions with other proteins. Importantly, recent experiments have provided direct evidence that histone acetylation facilitates the binding of transcription factors to nucleosomal DNA, indicating that histone acetylation increases the accessibility of chromatin to DNA-binding proteins. In addition, direct links between histone acetylation and transcriptional regulation have come from experiments showing that transcriptional activators and repressors are associated with histone acetyltransferases and deacetylases, respectively. This association was first revealed by cloning a gene encoding a histone acetyltransferase from Tetrahymena. Unexpectedly, the sequence of this histone acetyltransferase revealed that it was closely related to a previously known yeast transcriptional coactivator called Gcn5p, which stimulates transcription in association with several different sequence-specific transcriptional activators. Further experiments revealed that Gcn5p itself has histone acetyltransferase activity, suggesting that transcriptional activation results directly from histone acetylation. These results have been extended by the finding that several mammalian transcriptional coactivators are also histone acetyltransferases, as is a general transcription factor (TAFII250, a component of TFIID). Conversely, histone deacetylases (which remove the acetyl groups from histone tails) are associated with transcriptional repressors in both yeast and mammalian cells. Histone acetylation is thus regulated by both transcriptional activators and repressors, indicating that it plays a key role in eukaryotic gene expression.

(A) The core histones have histone-fold domains, which interact with other histones and with DNA in the nucleosome, and N-terminal tails, which extend outside of the nucleosome. The N-terminal tails of the core histones (e.g., H3) are modified by the addition of acetyl groups (Ac) to the side chains of specific lysine residues. (B) Transcriptional activators and repressors are associated with coactivators and corepressors, which have histone acetyltransferase (HAT) and histone deacetylase (HDAC) activities, respectively. Histone acetylation is characteristic of actively transcribed chromatin and may weaken the binding of histones to DNA or alter their interactions with other proteins.

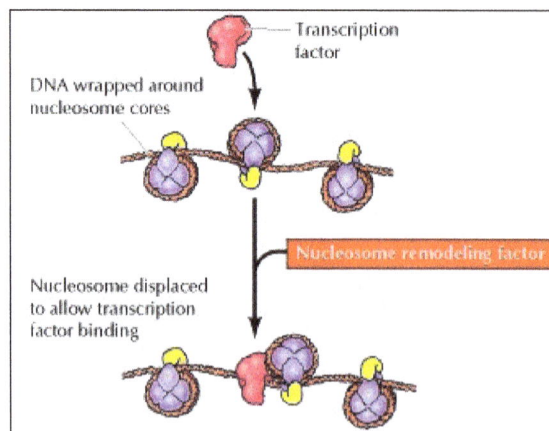

Figure: Nucleosome remodeling factors.

Nucleosome remodeling factors are protein complexes that facilitate the binding of transcription factors by altering nucleosome structure. The mechanism of action of nucleosome remodeling factors is not yet clear, but they appear to increase the accessibility of nucleosomal DNA to other proteins (such as transcription factors) without removing the histones. One possibility is that they catalyze the sliding of histone octamers along the DNA molecule, thereby repositioning nucleosomes to facilitate transcription factor binding. The mechanisms by which nucleosome remodeling factors are targeted to actively transcribed genes also remain to be established, although some studies suggest that they can be brought to enhancer or promoter sites in association with transcriptional activators or as components of the RNA polymerase II holoenzyme.

Nucleosome remodeling factors facilitate the binding of transcription factors to chromatin by repositioning nucleosomes on the DNA.

Perhaps surprisingly, the packaging of DNA in nucleosomes does not present an impassable barrier to transcriptional elongation by RNA polymerase, which is able to transcribe through a nucleosome core by disrupting histone-DNA contacts. The ability of RNA polymerase to transcribe chromatin templates is facilitated by acetylation of histones and by the association of the nonhistone chromosomal proteins HMG-14 and HMG-17 with the nucleosomes of actively transcribed genes. The binding sites of these proteins on nucleosomes overlap the binding site of histone H1, and HMG-14 and HMG-17 appear to stimulate transcription by altering the interaction of histone H1 with nucleosomes to maintain a decondensed chromatin structure that facilitates transcription through a nucleosome template. As with nucleosome remodeling factors, the signals that target HMG-14 and HMG-17 to actively transcribed genes remain to be elucidated by future research.

## DNA Methylation

The methylation of DNA is another general mechanism by which control of transcription in vertebrates is linked to chromatin structure. Cytosine residues in vertebrate DNA can be modified by the addition of methyl groups at the 5-carbon position. DNA is methylated specifically at the C's that precede G's in the DNA chain (CpG dinucleotides). This methylation is correlated with reduced transcriptional activity of genes that contain high frequencies of CpG dinucleotides in the vicinity of their promoters. Methylation inhibits transcription of these genes via the action of a protein, MeCP2 that specifically binds to methylated DNA and represses transcription. Interestingly, MeCP2 functions as a complex with histone deacetylase, linking DNA methylation to alterations in histone acetylation and nucleosome structure.

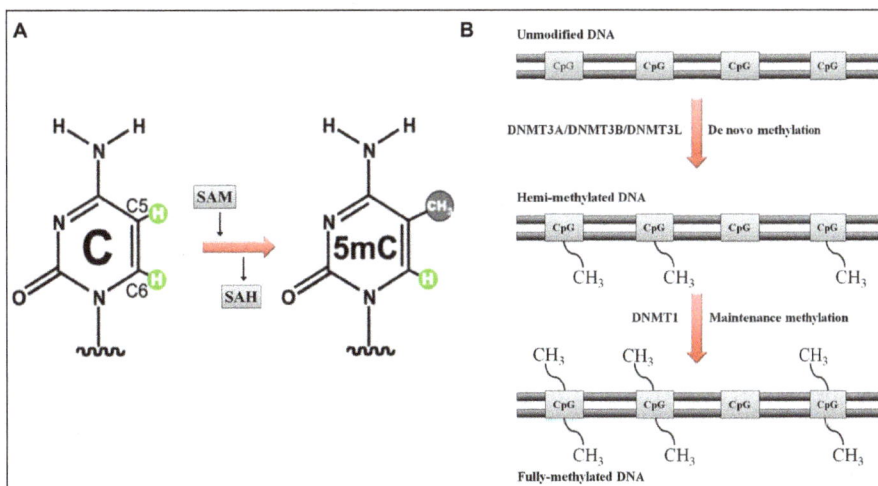

Figure: DNA methylation - A methyl group is added to the 5-carbon position of cytosine residues in DNA.

Although DNA methylation is capable of inhibiting transcription, its general significance in gene regulation is unclear. In many cases, methylation of inactive genes is thought to be a consequence, rather than the primary cause, of their lack of transcriptional activity. However, an important regulatory role of DNA methylation has been established in the phenomenon known as genomic imprinting, which controls the expression of some genes involved in the development of mammalian embryos. In most cases, both the paternal and maternal alleles of a gene are expressed in diploid cells. However, there are a few imprinted genes (over two dozen have been described in mice and humans) whose expression depends on whether they are inherited from the mother or from the father. In some cases, only the paternal allele of an imprinted gene is expressed, and the maternal allele is transcriptionally inactive. For other imprinted genes, the maternal allele is expressed and the paternal allele is inactive.

Although the biological role of genomic imprinting is uncertain, DNA methylation appears to distinguish between the paternal and maternal alleles of imprinted genes. A good example is the gene H19, which is transcribed only from the maternal copy. The H19 gene is specifically methylated during the development of male, but not female, germ cells. The union of sperm and egg at fertilization therefore yields an embryo containing a methylated paternal allele and an unmethylated maternal allele of the gene. These differences in methylation are maintained following DNA replication by an enzyme that specifically methylates CpG sequences of a daughter strand that is hydrogen-bonded to a methylated parental strand. The paternal H19 allele therefore remains methylated, and transcriptionally inactive, in embryonic cells and somatic tissues. However, the paternal H19 allele becomes demethylated in the germ line, allowing a new pattern of methylation to be established for transmittal to the next generation.

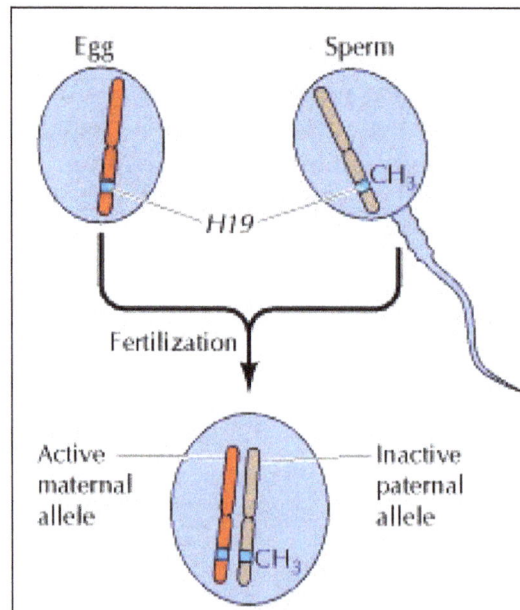

Figure: Genomic imprinting.

The H19 gene is specifically methylated during development of male germ cells. Therefore, sperm contain a methylated H19 allele and eggs contain an unmethylated allele. Following fertilization, the methylated paternal allele remains transcriptionally inactive, and only the unmethylated maternal allele is expressed in the embryo.

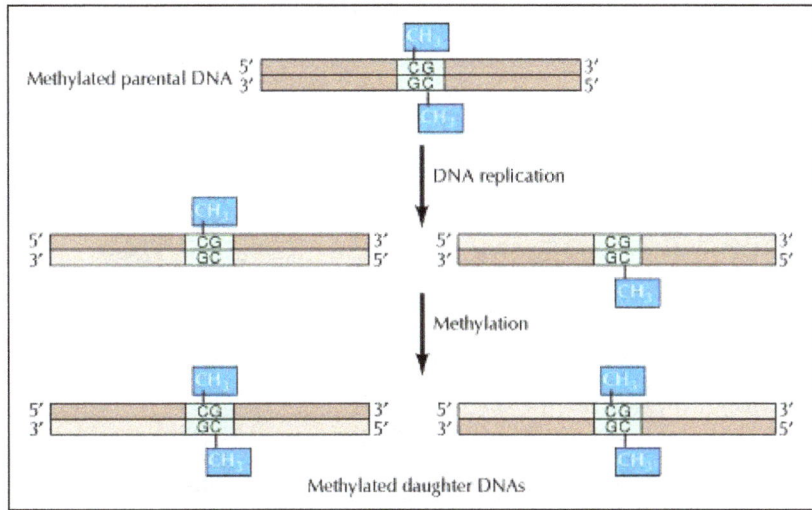

Figure: Maintenance of methylation patterns.

In parental DNA, both strands are methylated at complementary CpG sequences. Following replication, only the parental strand of each daughter molecule is methylated. The newly synthesized daughter strands are then methylated by an enzyme that specifically recognizes CpG sequences opposite a methylation site.

# RNA Processing

## mRNA Processing

Eukaryotic pre-mRNA receives a 5′ cap and a 3′ poly (A) tail before introns are removed and the mRNA is considered ready for translation.

## Pre-mRNA Processing

The eukaryotic pre-mRNA undergoes extensive processing before it is ready to be translated. The additional steps involved in eukaryotic mRNA maturation create a molecule with a much longer half-life than a prokaryotic mRNA. Eukaryotic mRNAs last for several hours, whereas the typical E. coli mRNA lasts no more than five seconds.

Pre-mRNAs are first coated in RNA-stabilizing proteins; these protect the pre-mRNA from degradation while it is processed and exported out of the nucleus. The three most important steps of pre-mRNA processing are the addition of stabilizing and signaling factors at the 5′ and 3′ ends of the molecule, and the removal of intervening sequences that do not specify the appropriate amino acids. In rare cases, the mRNA transcript can be "edited" after it is transcribed.

## 5′ Capping

While the pre-mRNA is still being synthesized, a 7-methylguanosine cap is added to the 5′ end of the growing transcript by a 5′-to-5′ phosphate linkage. This moiety protects the nascent mRNA from degradation. In addition, initiation factors involved in protein synthesis recognize the cap to help initiate translation by ribosomes.

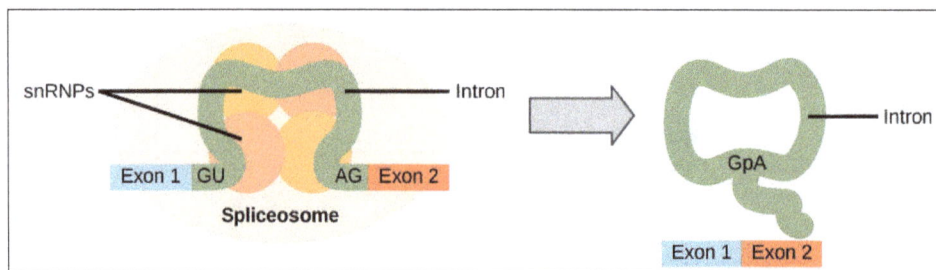

5′ cap structure: Capping of the pre-mRNA involves the addition of 7-methylguanosine (m⁷G) to the 5′ end. The cap protects the 5′ end of the primary RNA transcript from attack by ribonucleases and is recognized by eukaryotic initiation factors involved in assembling the ribosome on the mature mRNA prior to initiating translation.

## 3′ Poly A Tail

While RNA Polymerase II is still transcribing downstream of the proper end of a gene, the pre-mRNA is cleaved by an endonuclease-containing protein complex between an AAUAAA consensus sequence and a GU-rich sequence. This releases the functional pre-mRNA from the rest of the transcript, which is still attached to the RNA Polymerase. An enzyme called poly (A) polymerase (PAP) is part of the same protein complex that cleaves the pre-mRNA and it immediately adds a string of approximately 200 A nucleotides, called the poly (A) tail, to the 3′ end of the just-cleaved pre-mRNA. The poly (A) tail protects the mRNA from degradation, aids in the export of the mature mRNA to the cytoplasm, and is involved in binding proteins involved in initiating translation.

## Pre-mRNA Splicing

Eukaryotic genes are composed of exons, which correspond to protein-coding sequences (ex-on signifies that they are expressed), and intervening sequences called introns (int-ron denotes their intervening role), which may be involved in gene regulation, but are removed from the pre-mRNA during processing. Intron sequences in mRNA do not encode functional proteins.

## Discovery of Introns

The discovery of introns came as a surprise to researchers in the 1970s who expected that pre-mRNAs would specify protein sequences without further processing, as they had observed in prokaryotes. The genes of higher eukaryotes very often contain one or more introns. While these regions may correspond to regulatory sequences, the biological significance of having many introns or having very long introns in a gene is unclear. It is possible that introns slow down gene expression because it takes longer to transcribe pre-mRNAs with lots of introns. Alternatively, introns may be non-functional sequence remnants left over from the fusion of ancient genes throughout evolution. This is supported by the fact that separate exons often encode separate protein subunits or domains. For the most part, the sequences of introns can be mutated without ultimately affecting the protein product.

## Intron Processing

All introns in a pre-mRNA must be completely and precisely removed before protein synthesis. If the process errs by even a single nucleotide, the reading frame of the rejoined exons would shift, and the resulting protein would be dysfunctional. The process of removing introns and reconnecting exons is called splicing. Introns are removed and degraded while the pre-mRNA is still in the nucleus. Splicing occurs by a sequence-specific mechanism that ensures introns will be removed and exons rejoined with the accuracy and precision of a single nucleotide. The splicing of pre-mRNAs is conducted by complexes of proteins and RNA molecules called spliceosomes.

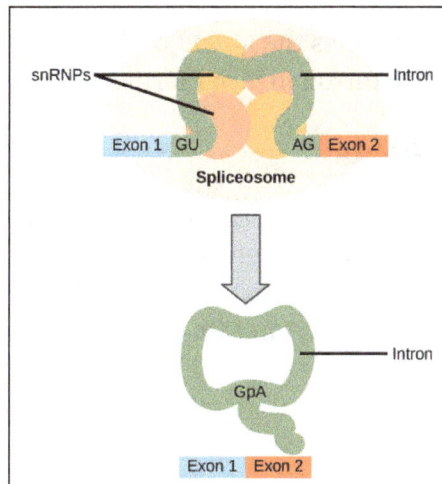

Pre-mRNA splicing: Pre-mRNA splicing involves the precise removal of introns from the primary RNA transcript. The splicing process is catalyzed by large complexes called spliceosomes. Each spliceosome is composed of five subunits called snRNPs. The spliceseome's actions result in the splicing together of the two exons and the release of the intron in a lariat form.

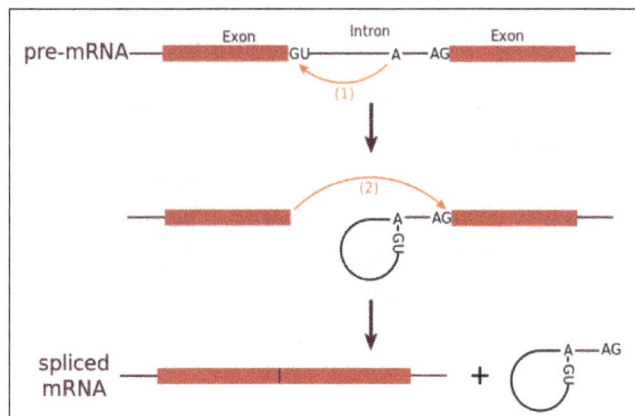

Mechanism of pre-mRNA splicing.

Each spliceosome is composed of five subunits called snRNPs (for small nuclear ribonucleoparticles, and pronounced "snurps"). Each snRNP is itself a complex of proteins and a special type of RNA found only in the nucleus called snRNAs (small nuclear RNAs). Spliceosomes recognize sequences at the 5' end of the intron because introns always start with the nucleotides GU and they recognize sequences at the 3' end of the intron because they always end with the nucleotides

AG. The spliceosome cleaves the pre-mRNA's sugar phosphate backbone at the G that starts the intron and then covalently attaches that G to an internal A nucleotide within the intron. Then the spliceosme connects the 3′ end of the first exon to the 5′ end of the following exon, cleaving the 3′ end of the intron in the process. This results in the splicing together of the two exons and the release of the intron in a lariat form.

The snRNPs of the spliceosome were left out of this figure, but it shows the sites within the intron whose interactions are catalyzed by the spliceosome. Initially, the conserved G which starts an intron is cleaved from the 3′ end of the exon upstream to it and the G is covalently attached to an internal A within the intron. Then the 3′ end of the just-released exon is joined to the 5′ end of the next exon, cleaving the bond that attaches the 3′ end of the intron to its adjacent exon. This both joins the two exons and removes the intron in lariat form.

## Processing of tRNAs and rRNAs

The tRNAs and rRNAs are structural molecules that have roles in protein synthesis; however, these RNAs are not themselves translated. In eukaryotes, pre-rRNAs are transcribed, processed, and assembled into ribosomes in the nucleolus, while pre-tRNAs are transcribed and processed in the nucleus and then released into the cytoplasm where they are linked to free amino acids for protein synthesis.

## Ribosomal RNA

The four rRNAs in eukaryotes are first transcribed as two long precursor molecules. One contains just the pre-rRNA that will be processed into the 5S rRNA; the other spans the 28S, 5.8S, and 18S rRNAs. Enzymes then cleave the precursors into subunits corresponding to each rRNA. In bacteria, there are only three rRNAs and all are transcribed in one long precursor molecule that is cleaved into the individual rRNAs. Some of the bases of pre-rRNAs are methylated for added stability. Mature rRNAs make up 50-60% of each ribosome. Some of a ribosome's RNA molecules are purely structural, whereas others have catalytic or binding activities.

The eukaryotic ribosome is composed of two subunits: a large subunit (60S) and a small subunit (40S). The 60S subunit is composed of the 28S rRNA, 5.8S rRNA, 5S rRNA, and 50 proteins. The 40S subunit is composed of the 18S rRNA and 33 proteins. The bacterial ribosome is composed of two similar subunits, with slightly different components. The bacterial large subunit is called the 50S subunit and is composed of the 23S rRNA, 5S rRNA, and 31 proteins, while the bacterial small subunit is called the 30S subunit and is composed of the 16S rRNA and 21 proteins.

The two subunits join to constitute a functioning ribosome that is capable of creating proteins.

## Transfer RNA

Each different tRNA binds to a specific amino acid and transfers it to the ribosome. Mature tRNAs take on a three-dimensional structure through intramolecular basepairing to position the amino acid binding site at one end and the anticodon in an unbasepaired loop of nucleotides at the other end. The anticodon is a three-nucleotide sequence, unique to each different tRNA that interacts with a messenger RNA (mRNA) codon through complementary base pairing.

There are different tRNAs for the 21 different amino acids. Most amino acids can be carried by more than one tRNA.

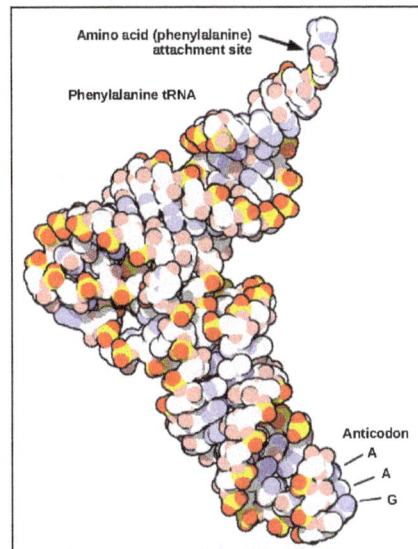

Structure of tRNA: This is a space-filling model of a tRNA molecule that adds the amino acid phenylalanine to a growing polypeptide chain. The anticodon AAG binds the codon UUC on the mRNA. The amino acid phenylalanine is attached to the other end of the tRNA.

In all organisms, tRNAs are transcribed in a pre-tRNA form that requires multiple processing steps before the mature tRNA is ready for use in translation. In bacteria, multiple tRNAs are often transcribed as a single RNA. The first step in their processing is the digestion of the RNA to release individual pre-tRNAs. In archaea and eukaryotes, each pre-tRNA is transcribed as a separate transcript.

The processing to convert the pre-tRNA to a mature tRNA involves five steps-

1. The 5′ end of the pre-tRNA, called the 5′ leader sequence, is cleaved off.

2. The 3′ end of the pre-tRNA is cleaved off.

3. In all eukaryote pre-tRNAs, but in only some bacterial and archaeal pre-tRNAs, a CCA sequence of nucleotides is added to the 3′ end of the pre-tRNA after the original 3′ end is trimmed off. Some bacteria and archaea pre-tRNAs already have the CCA encoded in their transcript immediately upstream of the 3′ cleavage site, so they don't need to add one. The CCA at the 3′ end of the mature tRNA will be the site at which the tRNA's amino acid will be added.

4. Multiple nucleotides in the pre-tRNA are chemically modified, altering their nitorgen bases. On average about 12 nucleotides are modified per tRNA. The most common modifications are the conversion of adenine (A) to pseudouridine (ψ), the conversion of adenine to inosine (I), and the conversion of uridine to dihydrouridine (D). But over 100 other modifications can occur.

5. A significant number of eukaryotic and archaeal pre-tRNAs have introns that have to be spliced out. Introns are rarer in bacterial pre-tRNAs, but do occur occasionally and are spliced out.

After processing, the mature pre-tRNA is ready to have its cognate amino acid attached. The cognate amino acid for a tRNA is the one specified by its anticodon. Attaching this amino acid is called charging the tRNA. In eukaryotes, the mature tRNA is generated in the nucleus, and then exported to the cytoplasm for charging.

# Post-transcriptional Gene Regulation

## RNA Localization

RNA localization generally refers to the transport or enrichment of subsets of mRNAs to specific subcellular regions. RNA localization can be achieved 'passively' by local protection from degradation or through the trapping/anchoring at specific cellular locations. Moreover, asymmetric distribution of RNA can also be established by the 'active' transport of RNAs via RBP-motor protein complexes.

In a pioneering study by Pat Brown and colleagues, mRNA species bound to 'membrane-associated' ribosomes were separated from free 'cytosolic' ribosomes by equilibrium density centrifugation in a sucrose gradient, and the distribution of transcripts in the fractions were quantified by comparative DNA microarray analysis. As expected, transcripts known to encode secreted or membrane-associated proteins were enriched in the membrane-bound fraction, whereas those known to encode cytoplasmic or nuclear proteins were preferentially enriched in the fractions containing mRNAs associated with cytoplasmic ribosomes. However, transcripts for more than 300 genes in the yeast Saccharomyces cerevisiae were found in the 'membrane-fraction' coding for previously unrecognized membrane or secreted proteins. Rather unexpected, among these was also the message for ASH1 coding for a well-known transcriptional repressor, suggesting alternative signals for membrane association. Similarly, application of this method to map mRNA distributions in the plant Arabidopsis thaliana allowed the classification of 300 previously unknown transcripts as secreted or membrane-associated proteins. A recent extension of this approach to eleven different human cell lines provided a detailed catalog containing more than 5000 previously uncharacterized membrane-associated and 6000 cytoplasmic/nuclear proteins at high confidence levels. Strikingly, this analysis predicts that 44% of all human genes encode membrane-associated or secreted proteins exceeding previous estimates ranging from 15% to 30%. In addition, the comparison of this catalog to data obtained from hundreds of DNA microarray profiles from tumors and normal tissues allowed the identification of candidate genes that are highly overexpressed in tumors and, hence, could be particularly good candidates for diagnostic tests or molecular therapies.

Claude Jacq's lab applied a subcellular fractionation approach to determine transcripts associated with free and mitochondrion-associated ribosomes in the yeast S. cerevisiae. Besides the mRNA for ATP14, which was previously known to localize in the vicinity of mitochondria, nuclear transcripts for diverse mitochondrial proteins were enriched in the mitochondrial fraction. Interestingly, two characteristics correlated with this mRNA localization: the phylogenetic origin and the length of the genes. mRNAs enriched in the mitochondrial fraction were preferentially longer (as deduced from average length of the encoded proteins) and originate from genes with bacterial homologues, whereas mRNAs in free cytosolic polysomes were shorter and of eukaryotic origin. Possibly, such coordinate localization of groups of mRNAs could allow oriented access for controlling their fates. This may also apply to other cellular compartments. For instance, a low-density array study revealed that 22 out of 649 analyzed transcripts were enriched in the cytoskeleton fraction relative to the cytosolic fraction — most of these encoding ribosomal proteins or structural proteins that interact with the cytoskeleton.

In polarized cells like neurons, mRNA localization has major physiological implications. In dendrites, RNA transport and subsequent local protein synthesis is thought to influence experience-based synaptic plasticity and long-term memory formation; in axons, local translation modifies axon guidance and synapse formation. However, to date there are only a handful of well-characterized examples of localized neuronal mRNAs, among them the messages coding for microtubule-associated protein 2 (MAP2), the α-subunit of a calmodulin-dependent protein kinase (αCaMKII), brain-derived neurotrophic factor (BDNF), and activity-regulated cytoskeletal-related (Arc). Therefore, several genomics-based approaches have been undertaken to identify novel localized transcripts. For example, Matsumoto et al. fractionated brain tissue and isolated RNA from the heavy portion of polysomes and synaptosomes to provide a list of potentially dendritic mRNAs that undergo localized translation. Interestingly, the induction of neural activity by an electroconvulsive shock triggered a redistribution of the population of dendritic transcriptome, which may trigger changes in the translatability of this transcriptome, suggesting complex mechanisms of local translation in response to synaptic inputs. The hundreds of potentially localized mRNA in neurons now await confirmation by in situ hybridization and exploration as to whether and how these may be regulated through activating stimuli, such as neurotransmitter release.

To date, more than 100 mRNAs are known to undergo active mRNA transport in diverse organisms. In neurons, mRNAs are transported over long distances in a microtubule-dependent manner in the form of large granules consisting of RNA-binding proteins, ribosomes and translation factors. Several RBPs associated with neuronal RNA transport have been identified, such as zipcode-binding proteins (ZBP1,2; named after their ability to bind to a conserved 54-nucleotide element in the 3'-UTR of the β-actin mRNA known as the 'zipcode'), Staufen, hnRNPA2, cytoplasmic polyadenylation protein (CPEB) and members of the familial mental retardation proteins (FMRPs). At least for one of these RBPs, FMRP, a systematic gene array-based screen was undertaken to identify the mRNAs that are transported and possibly regulated by this protein. Using a 'ribonomics' approach, which involved the immunoprecipitation or affinity isolation of RBPs followed by the identification of bound RNAs with DNA microarrays, the scinetists immunopurified the protein from mouse brain tissues and found ~4% of all mRNAs (435 messages) associated with FMRP. In addition, they compared the mRNA profiles of polyribosomes between normal human cells and cells derived from fragile X syndrome patients identifying over 200 messages with altered association and hence, these are potentially subject to translational regulation. Notably, nearly 70% of the homologous messages found in both studies had a G-quartet structure, which was demonstrated as an in vitro FMRP target. These data provided a good starting point for further investigations on the most critical targets involved in fragile X pathophysiology and, possibly, on other related cognitive diseases.

Probably the best-studied example for actin-dependent RNA transport concerns ASH1 mRNA localization to the bud tip of yeast cells during cell division. ASH1 codes for a transcriptional repressor repressing mating type switching in daughter cells. ASH1 mRNA is bound by She2p, an RBP tethered to the myosin motor protein Myo4p via the adaptor protein She3p. This RNA-protein complex travels along actin cables to the emerging bud for local protein synthesis. To identify other localized mRNAs, affinity purification of components of the She complex followed by the analysis of bound mRNA with microarrays was combined with a robust reporter system for in vivo visualization as a secondary screen. This analysis revealed 23 additional transcripts that are localized to the bud-tip and encode a wide variety of proteins, several involved in stress responses and cell wall maintenance. These results reveal an unanticipated widespread use of RNA transport in budding yeast — possibly providing the daughter cells with a favorable 'start-up package'.

The few studies that investigated spatial distribution of mRNAs in the cell on a global level challenge the long-standing assumption of a rather 'unorganized' pool of mRNAs that randomly diffuse in the cytoplasm to be eventually translated. Possibly, many mRNAs may be spatially organized even in non-polarized cells for local translation or decay in processing (P) bodies. Further applications of both subcellular fraction techniques and ribonomics approaches will certainly reveal a more comprehensive picture of the spatial arrangement of RNAs in cells.

## Regulation of Translation

Translational regulation has essential roles in development, oncogenesis and synaptic plasticity. It concerns the differential recruitment of mRNA species to the ribosome for protein synthesis, which results in a lack of correlation between the relative amounts of mRNA and the amount of the encoded protein. In an innovative study, the relative contribution of transcriptional and translational regulation in yeast was measured using large-scale absolute protein expression measurement called APEX (Absolute Protein Expression Index), which relies upon observed peptide counts from mass spectrometry. Most (73%) of the variance in protein abundance can be explained by mRNA abundance, which is lognormal distributed around an average of 5600 proteins per mRNA molecule. This indicates that the abundance of most proteins is set per mRNA molecule; however, one third of the mRNAs must be regulated at additional levels including translation and protein turnover. In mammalian cells, the fraction of differentially expressed messages may be considerably higher ranging from 60% to 80%, indicating that gene expression of most messages is heavily controlled at diverse levels.

Translation can be divided into three steps: initiation, elongation and termination. During translation initiation, the primary target for translational control, translation initiation factors (eIFs) recruit the mRNA to the small ribosomal subunit (40S subunit). Thereby, eIF4E binds to the cap structure at the 5'-end of the mRNA and interacts with eIF4G, which binds to the poly (A)-binding protein (PABP). The initiation complex then scans the mRNA in 5' to 3' direction until the initiation codon is reached where the large ribosomal subunit (60S) joins the complex leading to the formation of active ribosomes. Notably, ribosomes can also be recruited cap-independently to some viral and cellular mRNAs by direct binding of the small ribosomal subunit to internal RNA structures, termed IRES. The assembled ribosomes traverse the coding region with help of elongation factors (eEFs) and synthesize the encoded polypeptide with multiple ribosomes covering them RNA to form polysomes. At the termination codon, peptide chain-releasing factors (eRFs) are required to release the polypeptide from the ribosome.

Two basic modes of translation regulation have been described. During global regulation, translation of most mRNAs is controlled by translation factors. For instance, phosphorylation of eIF2α reduces the amount of active initiation complexes and hence leads to a rapid reduction of translation of most messages. The availability of eIF4E is controlled by 4E-binding proteins (4E-BP) that displace eIF4G from eIF4E, and thus inhibit association of the small ribosomal subunit with them RNA. The second mode of translational regulation concerns mRNA-specific control, where translation of defined groups of mRNAs is modulated without affecting general protein biosynthesis. This can be carried out by specific RNA-binding proteins, which often bind to sequence or structural elements in untranslated regions (UTRs) of protein-coding transcripts and, hence, repress translation via interactions with eIFs. A prime example for such regulation represents cytoplasmic aconitase, an enzyme that regulates iron-dependent translation initiation through binding to a stem-loop structure in the 5'-UTR of messages involved in iron metabolism (e.g., ferritin mRNA coding for an iron regulatory protein). Specific control can also be exerted by microRNAs (miRNAs) — small RNAs of 22 nucleotides in length — that have recently been shown to repress translation via base pairing to sequences located in 3'-UTRs of target mRNAs. Interestingly, it has recently become apparent that miRNA- and RBP-mediated translational regulation may collaborate or compete on specific mRNA substrates, suggesting interconnections between these different modes of translational regulation.

## Genome-wide Analysis of Translational Regulation

A reliable measure for translation of cellular mRNA is the degree of its association with ribosomes. Since the rate of initiation usually limits translation, most translational responses will alter the ribosome density on a given mRNA. Actively translated mRNAs are typically bound by several ribosomes (polysomes) and can be separated from the small (40S) and the large (60S) ribosomal subunits and the 80S monosomes by sucrose gradient centrifugation. In classical experiments, total RNA was isolated from fractions of the polysomal gradient and assayed for the mRNA of interest by Northern blot analysis. Several laboratories have further extended this technique using DNA microarray technology to perform genome-wide analysis of mRNAs in polysomes in yeast, Drosophila and mammals.

The laboratories of Pat Brown and Daniel Herschlag performed a high resolution translation state analysis in rapidly growing yeast cells, providing profiles for mRNA-ribosome association for thousands of genes. Based on these data, they calculated the ribosome occupancy (fraction of a specific mRNA associated with ribosomes), the ribosome density and the translation rate for each expressed mRNA. The average occupancy was calculated at 71%, indicating that most mRNAs are likely engaged in active translation. However, about 100 mRNAs showed only weak association with ribosomes and may therefore be considered as potential candidates for 'translation on demand'. The average ribosome density was found to be 156 nucleotides per ribosome, which is about one fifth of the maximal packing density, supporting the premise that translation initiation is the rate-limiting step for protein synthesis. Surprisingly, the ORF length appears to be a major factor determining the ribosome density, which is expressed through an inverse correlation between the ORF length and the ribosome density in diverse species.

Since poly(A) tail length affects translational efficiencies and mRNA stability, two recent studies systematically addressed its length in yeast. In a procedure called polyadenylation state array analysis (PASTA), mRNAs were captured with poly(U) Sepharose columns and differentially eluted by increasing temperature. The mRNAs with short tails elute first and those with long tails last. RNAs fractions were analyzed with DNA microarrays to identify groups of mRNAs with similar poly(A) tail lengths. In the yeast S. cerevisiae, mRNA coding for functionally or cytotopically related mRNAs could be attributed to groups with similar tail length. Long poly(A) tails were found among mRNA coding for cytoplasmic ribosomal proteins, whereas short tails were enriched for DNA/Ty elements and among mRNAs coding for nucleolar proteins involved in ribosome synthesis, and proteins with cell cycle-related functions. The comparison of the data with other genome-wide analysis revealed that poly(A) tail length positively correlates with ribosome density and to some extend with mRNA abundance, and it negatively correlates with ORF and UTR length. The poly(A) tail length, and hence ribosome occupancy of messages correlate with the degree of association with poly(A)-binding protein (Pab1p). This provides 'global' support for the concept that long poly(A) tails stimulate translation via Pab1p and eIF4G. Interestingly, poly(A) tail length does not correlate with mRNA decay rates. Therefore, it appears that translation rates are not directly coupled to mRNA decay control, although poly(A) shortening is a prerequisite for mRNA decay. Possibly, processes acting on oligo(A)-tailed intermediates may limit the decay rates of large number of yeast mRNAs. A congruent study performed in fission yeast S. pombe monitored translational status, poly(A) tail length, mRNA abundance, mRNA decay rates and RNA polymerase II association under identical conditions. Functional groupings of mRNA in respect to their translational efficiencies, length and abundance were identified with shorter and abundant mRNAs having longer poly(A) tails. Notably, ORF length correlated best with ribosome density and mRNA abundance with ribosome occupancy. Both studies revealed similar principles that may organize translation and therefore, these may be evolutionarily conserved. Further studies in other organisms will reveal whether these principles are universally conserved and possibly affected in disease.

Several studies were aimed at the systematic identification of translationally regulated messages after subjecting cells to stress and other environmental stimuli. They applied a 'low-resolution' profile analysis, where the mRNA contents of high sucrose gradient fractions (polysomes) were compared with fractions from the low sucrose gradient (the pool of non-translated mRNAs). In parallel, changes in the levels of total RNA were measured to study the relation between transcription/decay and translationally regulated messages. In yeast, global effects on translation were first studied for the rapid transfer of cells from a fermentable to a non-fermentable carbon source mimicking glucose starvation, followed by heat-shock response and rapamycin treatment, amino acid starvation, butanol addition (an end product of amino acid breakdown), and application of hydrogen peroxide to induce oxidative stress. First, it should be noted that these relatively 'harsh' treatments induced global translation inhibition that goes along with a decrease in cell growth. Although this global translation inhibition is triggered by similar signaling pathways like phosphorylation of eIF2α, the various forms of stress affected quite different sets of mRNAs. Amino acid and glucose starvation differentially regulated the translation of up to 20% of all mRNAs, whereas more 'mild' treatments, such as butanol and peroxides, affected less than 4% of transcripts. Treatment of cells with heat/rapamycin or nutrient removal (amino acid and glucose starvation) co-activated similar translational and transcriptional programs. Here, regulation at the translational level often reflects a magnification of the transcriptional activity — an effect that has been termed 'potentiation'. In contrast, the addition of butanol or peroxide provoked no potentiation, but instead changed the abundance and translation rates of different sets of mRNAs. This could, at least in part, be explained by the recruitment of stored mRNA for translation 'on demand'. Nevertheless, in all cases, the specific sets of mRNAs that undergo treatment-specific regulation appear to share functional themes that can be attributed to logic response of the cell's altered physiological circumstances. For instance, mRNAs coding for proteins related to sugar metabolism and transport, such as hexose transporters, remain associated with polysomes during glucose starvation; rapamycin treatment, which blocks the target of rapamycin (TOR) pathway controlling cell growth, led to a decrease of nearly all yeast mRNAs coding for cytoplasmic ribosomal proteins, whereas mRNAs for proteins acting in the nitrogen discrimination pathway were increased; amino acid starvation strongly coregulated or potentiated transcripts encoding permeases, proteases and proteins involved in degradation pathways, which may reflect an early amino acid scavenging response to starvation. Another interesting aspect of these studies is that the concentration of an applied compound may significantly matter for the outcome. Shenton et al. showed that low concentrations of peroxide (0.2 m M $H_2O_2$) induced the translation of mRNAs coding for antioxidants, cellular transporters and proteins involved in diverse intermediary metabolism and may reflect the need for metabolic reconfiguration. A tenfold higher concentration of peroxide (2 mM) resulted in the up-regulation of genes involved in ribosome biogenesis and ribosomal RNA processing, possibly reflecting the need to repair factors for efficient protein synthesis.

On the other hand, many translationally repressed mRNAs showed increased steady-state (total RNA) levels. Again it was postulated that this group of messages may represent an mRNA store that could become rapidly activated following relief of the stress condition. It will certainly be interesting to further study whether other 'mild' treatments with pharmacological compounds activate dose-dependent non-linear effects via distinct regulatory programs. If so, this may become of great medical relevance as diverse drugs are known to act differentially depending on their dose.

Finally, there are recurring and intriguing observations that mRNAs coding for cytoplasmic ribosomes generally appear to undergo outstanding and strong coregulation. Amino acid and glucose starvation coordinately repress these transcript's abundances and ribosome association very rapidly, whereas the addition of butanol or oxidative stress even lead to the opposite effect — translational activation — that possibly reflects the requirement of cells to replace ribosomal proteins and rRNA that became damaged by free radicals or other toxic products. Therefore, besides tight transcriptional control of these messages, they also undergo decent post-transcriptional control at diverse levels and hence, may represent the most tightly controlled genes in eukaryotic cells.

First studies to investigate global aspects of translational regulation in mammalian cells focused on IRES-dependent translation in poliovirus-infected cells, and the reaction of mitogenically activated fibroblasts, providing early proof-of-principle for the methodology introduced above that involves polysomal fractionation followed by DNA microarray analysis of RNA contents. A further landmark study by Holland and colleagues analyzed polysomal profiles of murine cell lines after blocking oncogenic Ras and Akt signaling. Apparently, these pathways regulate the recruitment of specific mRNAs to ribosomes to a far greater extent than de novo synthesis of mRNAs by transcription and thus, Ras and Akt signaling pathways seem to have a more pronounced effect on translational versus transcriptional regulation. The authors postulated that the immediate and direct inductive oncogenic effect of these signaling pathways could be largely achieved through translational activation. The differences seen in RNA abundances during chronic signaling alterations may be secondary to translational effects caused by mRNAs encoding transcription factors. Similar studies on different cancer types may lead to the identification of potential markers and possibly reveal novel drug targets. Moreover, a recent study identified specific subsets of mRNAs regulated by eIF4E overexpression, which is known to lead to tumor transformation. The authors postulated that down-regulation of eIF4E and its downstream targets may represent a potential therapeutic option for the development of novel anti-cancer drug.

As seen for Ras/Akt activation, it is intriguing that changes at translation can even outperform changes at the steady-state mRNA level. This has also been noticed in a study analyzing radiation-induced changes in gene expression of human brain tumor cells or normal astrocytes. Ten times more genes (~15%) were altered at the level of translation compared to the number of genes regulated at the level of transcription (~1.5% of 7800 analyzed human genes). Only a few transcripts were commonly affected at both the transcriptional and translational levels, suggesting that the radiation-induced changes in transcription and translation are not coordinated. Those transcripts that were affected at translation fell into functional groups such as cell cycle, DNA replication and anti-apoptotic functions. This indicates that DNA damage affects post-transcriptional gene regulation of previously synthesized mRNAs, possibly enabling cells to repair DNA instead of being transcriptionally active. Functional relations among messages were also recognized in a recent study performing translational profiling of mouse pancreatic β-cells in response to an acute increase in glucose concentration. More than 300 transcripts (2% of the analyzed genes) changed their association with polysomes more than 1.5-fold; most of them encoding proteins acting in metabolism or transcription. Notably, this set of messages is related to the group of genes translationally altered during glucose starvation in the yeast S. cerevisia. Therefore, in mammals and yeast, it appears that mRNAs for functionally related messages may be coordinately regulated at the translational level. It is possible that a comparative analysis in different species may allow evolutionarily conserved translational regulatory programs to be deciphered, which are at the moment still rather speculative.

Whereas concomitant changes in RNA abundance and translational rates were rarely detected during radiation response, a recent study identifying mRNAs that remain associated with polysomes during hypoxia in transformed prostate cancer cells (a condition that tumors prevent through the induction of angiogenesis) found both homodirectionally/potentiated mRNAs and distinctively regulated messages. After prolonged exposure of PC-3 cells to low oxygen levels, global translation was reduced by half; however, 104 mRNAs, representing about 0.5% of all analyzed features, became more associated with hypoxic polysomes compared to normoxic ones. Among these, 71 mRNAs were similarly increased in hypoxic polysomes compared with total RNA levels representing homodirectional changes; 33 mRNAs were translationally enriched, some of them 'potentiated' (11 of those coding for ribosomal proteins).

The common principles of translational regulation that emerge from genome-scale studies in diverse eukaryotes suggest a complex but coordinate system of regulation. It must be triggered by a variety of factors that go well beyond the described pathways that influence global translation. In future, there will certainly be an increasing number of studies to decipher translational regulatory programs in cancer, neurogenesis and development. Intriguingly, despite the impact of translational regulation during development, only one recent study systematically investigated translational programs during early embryogenesis in the fruit fly Drosophila melanogaster. The mapping of translational programs in diverse species will likely reveal key regulatory networks and how these are affected in disease.

## Regulation of mRNA Decay

Steady-state mRNA levels are a result of both RNA synthesis and degradation that are dynamically controlled and can vary up to 100-fold during the cell cycle or cellular differentiation. In eubacteria like Escherichia coli, mRNAs are generally degraded by endonucleolytic cleavage, followed by 3'-to-5' exonucleolytic RNA decay through the so-called RNA degradasome consisting of ribonuclease E (RNAseE), 3'-exoribonuclease polynucleotide phosphorylase (PNPase), RNA helicase (RhlB) and enolase. In eukaryotes, most cytoplasmic mRNA degradation begins with shortening of the poly(A) tail by deadenylases followed by removal of the 5' cap structure by the decapping enzymes, Dcp1 and Dcp2. The decapped intermediates are then degraded either by an exonuclease (Xrn1p) in the 5' to 3' direction, or by the cytoplasmic exosome in the 3' to 5' direction. In addition, eukaryotes own specialized pathways that target mRNAs containing premature termination codons (nonsense-mediated decay pathway, NMD) that lack translational termination codons (non-stop decay pathway, NSD) or that bear stalled ribosomes (no-go decay). Degradation of specific mRNAs can also be initiated by endonucleolytic cleavage through sequence-specific endonucleases, or in response to miRNAs or siRNAs. Numerous cis-acting elements located in the 5'-UTR, the coding sequence (CDS) or in the 3'-UTR of mRNAs can function as binding sites for RNA-binding proteins that regulate decay. For instance, AU-rich elements (AREs), conserved sequences found in the 3'-UTR of nearly 5% of all human genes, interact with specific ARE-binding proteins that stabilize the RNA or promote mRNA degradation by recruiting the RNA decay machinery.

## Genome-wide Measurements of mRNA decay

Global analysis of mRNA decay rates following arrest of transcription has been performed in all three kingdoms of life: bacteria, archea and eukaryotes including yeast, plants, and human cells.

In eubacteria and archea, mRNA decay proceeds rapidly, with a median half-life of ~5 min. Two main characteristics seem to be evolutionarily conserved: adjusted decay rates for functionally related groups of messages, and the inverse correlation between the half-lives and the relative abundances of transcripts. As seen in the archaebacterium Sulfolobus, transcripts encoding proteins involved in growth-related processes, such as transcription, tRNA synthesis, translation and energy production, generally decay rapidly ($t_{1/2} \leq 4$ min), whereas those encoding products necessary for maintaining cellular homeostasis are relatively stable ($t_{1/2} > 9$ min). Short half-lives of highly abundant mRNAs imply high-turnover rates and thus, enable cells to rapidly reprogram gene expression upon changes in environmental conditions. Interestingly, the half-life and abundance of distinct classes of transcripts appear to depend on particular RNA degradosome components. This finding suggests the existence of structural features or biochemical factors that distinguish different classes of mRNA targeted for degradation. This may also apply to specific growth phases, as seen in Streptococcus where certain mRNAs become sensitive to stationary-phase-induced PNPase.

Evidence for the existence of coordinated RNA decay regulons in eukaryotes was obtained from global investigation of mRNA decay profiles in yeast and human cells. Here, transcription was shut-off using cells that bear a temperature-sensitive allele of RNA polymerase II or through chemical inactivation, and the decay of thousands of genes was monitored with DNA microarrays over a time course. Strikingly, mRNA half-lives among components of macromolecular complexes in yeast were significantly correlated. For instance, the transcripts for the four histone mRNAs were among the least stable with closely matched, rapid decay rates ($t_{1/2}$=7±2 min); the 131 mRNAs coding for ribosomal proteins had average decay rates ($t_{1/2}$=22±6 min), and the four components of the trehalose phosphate synthetase complex were amongst the longest lived messages ($t_{1/2}$=105±15 min). The examination of decay rates in human cells revealed similar mRNA-turnover patterns among orthologous genes, indicating the presence of evolutionary conserved programs of RNA stability control. Transcripts encoding metabolic proteins have a tendency for longer half-lives, whereas transcripts encoding transcription factors or ribosome biogenesis factors are relatively short lived. Interestingly, it appears to be a universal feature that average transcript half-lives are roughly proportional to the length of the cell cycle: cell-cycle lengths of 20, 90, and 600 min correspond to median mRNA half-lives of 5, 21 and 600 min for E. coli, S. cerevisiae and human cells, respectively.

DNA microarrays have also been applied to investigate specialized decay pathways, such as NMD and nuclear exosome-mediated decay. Mutants for NMD factors Upf1, Nmd2 and Upf3 alter the mRNA levels of an overlapping set of ~600 messages (10% of the transcriptome) in yeast. However, mRNA levels in nmd⁻ strains may also be the result of indirect effects because transcription factors are also targeted through NMD and therefore, Guan et al. dissected direct from indirect targets of NMD by profiling global RNA decay rates in nmd⁻ strains. About half (300 transcripts) are likely to be direct NMD targets decayed through 5' to 3' degradation by Xrn1p. NMD-sensitive transcripts tend to be both non-abundant and short-lived, with one third of them coding for proteins that are connected to two central themes: first, replication and maintenance of telomeres, chromatin-mediated silencing and post-replication events related to the transmission of chromosomes during the cell division cycle; and, second, synthesis and breakdown of plasma membrane components, including transport of macromolecules and nutrients, and cell wall proteins. Genome-wide analyses have also identified potential RNA substrates for the nuclear exosome. More than 300 mRNAs showed altered expression levels in different exosome mutants. Several genes, located downstream of independently transcribed snoRNA genes, were overexpressed in exosome mutants. Further analyses suggested that many snoRNA and snRNA genes are inefficiently terminated. Such read-through transcripts into downstream ORFs are normally rapidly degraded by the exosome and, hence, could explain their enrichment in exosome mutants.

A couple of studies investigated the implications of specific RBPs on RNA turnover. Global mRNA turnover in mutant cells was monitored through gene expression analysis expecting adverse effects on subsets of messages. Grigull et al. examined the effects of deletions of genes encoding deadenylase components Ccr4p and Pan2p and putative RNA-binding proteins Pub1p and Puf4p after inhibition of transcription by chemicals and heat stress. This examination showed that Ccr4p, the major yeast mRNA deadenylase, contributes to the degradation of transcripts encoding both ribosomal proteins/rRNA synthesis and ribosome assembly factors largely mediating the transcriptional response to heat stress. Pan2p and Puf4p also participate in degradation of these mRNAs, while Pub1p preferentially stabilized transcripts encoding ribosomal proteins. Notably, the Puf4-affected genes correlate with biochemically identified targets of Puf4p. A second study focused on Pub1p, a yeast RNA-binding protein thought to destabilize mRNAs through binding to AU-rich sequences in 3'-UTRs. Global decay profiles in pub1 mutants revealed a significant destabilization of proteins involved in ribosomal biogenesis and cellular metabolism, whereas genes involved in transporter activity demonstrated association with the protein, but displayed no measurable changes in transcript stability. Therefore, in this case, the direct targets only partially related to the functional outcome under specific physiological conditions. This could be mediated through additional RNA protein interactions forming a network through combinatorial binding. Finally, Foat et al. combined a computational and experimental approach to identify transcripts that are destabilized under specific environmental conditions (sugar sources) by yeast mRNA stability regulators. For Puf3p, which was known to primarily associate with mRNAs coding for mitochondrial proteins, they computationally inferred and experimentally verified target destabilization in the presence of glucose, as some of these mRNAs were up-regulated in puf3 mutants grown in a non-repressing carbon source, but down-regulated in a repressing carbon source.

Mammalian cells have evolved a variety of specific mRNA decay programs that play important roles in medically relevant processes such as inflammation, hypoxia and cancer pathogenesis. For example, the expression of diverse cytokines is differentially regulated after T cell activation, and glucocorticoids inhibit inflammation through destabilization of proinflammatory transcripts like cyclooxygenase-2. Global mRNA decay profiles revealed mRNAs, which appear specifically regulated by these programs. For instance, in resting T lymphocytes, the majority of transcripts are stable with half-lives of more than 6 h, but a small proportion (~3%) of expressed transcripts exhibits rapid decay with half-lives of less than 45 min. These short-lived transcripts encode a variety of regulatory proteins such as cell surface receptors, transcription factors and regulators of cell growth and apoptosis. Su et al. focused on the massive degradation of transcripts occurring during meiotic arrest at the germinal-vesicle (GV) stage, and found that degradation is apparently not promiscuous but preferentially affects specific groups of messages. In particular, transcripts involved in processes associated with meiotic arrest at the GV stage and the progression of oocyte maturation, such oxidative phosphorylation, energy production, and protein synthesis, were rapidly degraded, whereas those encoding participants in signaling pathways maintaining the oocyte in the MII-arrested state were among the most stable. These studies exemplify that stimulus-dependent transcript destabilization is an important mechanisms for controlling gene expression in a coordinated manner.

Many activation-induced transcripts contain AREs in the 3'-UTR. The presence of these motifs in mRNAs often correlates with shifts in the distribution of decay rates; however, their sole presence cannot reliably predict turnover behavior. ARE-binding proteins may therefore differentially determine the fate of mRNA depending on the cellular and environmental context. Tristetraprolin (TTP), a well-known ARE-binding protein, has several characterized physiological target mRNAs including tumor necrosis factor (TNF)-α, granulocyte-macrophage colony-stimulating factor, and interleukin-2β. Micro-array analysis of RNA obtained from wild-type and TTP-deficient fibroblast cell lines identified 250 transcripts with altered decay rates, some of them containing conserved TTP binding sites. The RNA-binding protein T-cell intracellular antigen 1 (TIA-1) functions as a post-transcriptional regulator of gene expression and aggregates to form stress granules following cellular damage. TIA-1 regulates mRNAs for proteins involved in inflammatory responses such as TNF-α and cyclooxygenase 2. Immunoprecipitation (IP) of TIA-1-RNA complexes, followed by microarray-based identification and computational analysis of bound transcripts revealed at least 300 potential targets, many of them bearing a U-rich motif.

global analysis of mRNA turnover underlines the importance of RNA decay in the control of mRNA levels and strongly suggests the presence of specific RNA turnover programs. mRNA decay certainly involves combinatorial interactions of RBP enabling stimulus-dependent decay programs through the integration of diverse signals. Besides temporal control, RNA decay may also occur spatially restricted, as seen with *Drosophila* IRE1, a protein activated during the unfolded protein response in the endoplasmic reticulum directing the decay of specific subset of mRNAs, many of which encode plasma-membrane proteins. Moreover, still rather unexplored is the role of P-bodies and stress granules as storage place for untranslated mRNA and site for mRNA degradation. Perhaps different subtypes of P-bodies exist for subgroups of RNAs? At least the recent observation that ARE containing mRNAs are localized to specific cytoplasmic granular structures containing exosome subunits that are distinct from P-bodies or stress granules, support the idea of specialized structures for storage or degradation of distinct groups of mRNAs.

## Identification of Specific RNA-protein Interactions

Putative RNA-binding proteins comprise 3 –11% of the proteomes in bacteria, archea and eukaryotes. The large number of RBPs in all kingdoms of life may merely reflect the ancient origin of RNA regulation, which is possibly the most evolutionary conserved part of cell physiology. RBPs often contain distinct RNA-binding domains that specifically interact with sequences or structural elements in the RNA. Approximately one hundred protein domains associated with RNA metabolism have been described to date, half of them believed to have originated at early stages in evolution, whereas others, such as the RNA recognition motif (RRM), are exclusively present in eukaryotes and therefore may have been acquired later in evolution.

A successful approach to globally identify the in vivo RNA targets of RBPs involves immunoprecipitation or affinity purification of epitope-tagged proteins followed by the analysis of associated RNAs with DNA microarrays or by sequencing. In a pioneering study, Keene and colleagues used this 'ribonomics' approach to study RNAs associated with three RBPs in a cancer cell line. Although low-density arrays were used to identify the bound mRNAs, each RBP was associated with a distinct subset of the mRNAs present in total cell lysate. Moreover, these subsets appeared to change after cells were induced to differentiate. These results led to the proposal that groups of mRNAs encoding functionally related proteins are organized as so-called 'post-transcriptional operons'. In analogy to prokaryotic operons, this model predicts that specific RBPs may coordinate groups of mRNAs coding for functionally related proteins in eukaryotes. Cis-acting elements in the mRNA may provide the means to mimic the coordinated regulatory advantages of clustering genes into polycistronic operons.

A prime example for the coordination of functional related transcripts by specific RBPs is represented by the Pumilio-Fem-3 binding (PUF) proteins. PUF proteins comprise a conserved family of structurally related RBPs that negatively regulate gene expression of specific mRNAs. Applying DNA microarrays to identify their RNA targets revealed that each of the five yeast PUF proteins associated with distinct groups of 40 to 220 different mRNAs with striking common themes in the functions and subcellular localization of the proteins they encode: Puf3p binds nearly exclusively to cytoplasmic mRNAs that encode mitochondrial proteins; Puf1p and Puf2p interact preferentially with mRNAs encoding membrane-associated proteins; Puf4p preferentially binds mRNAs encoding nucleolar ribosomal RNA-processing factors; and Puf5p is associated with mRNAs encoding chromatin modifiers and components of the spindle pole body. The results were further corroborated by the identification of distinct sequence motifs in the 3′-untranslated regions of the mRNAs bound by Puf3, Puf4, and Puf5 proteins. A physiological relation between Puf3p and its mRNA targets has also been observed — as suggested from its association with mRNA-encoding mitochondrial proteins, Puf3 mutant cells showed a slow-growth phenotype on non-fermentable carbon sources indicative of a functional connection to mitochondrial physiology.

Genome-wide identification of RNAs associated with the orthologous PUF protein from Drosophila melanogaster, called PUMILIO, revealed distinct clusters of mRNAs in embryos and in ovaries of adult flies. More than 1000 messages were significantly associated with the protein. Subgroups of these Pum-associated mRNAs had commonalities, such as function in the anterior-posterior patterning system, and the subunits of the vacuolar H-ATPase. Moreover, a characteristic sequence motif was present in 3'-UTRs of PUMILIO-bound mRNAs resembling the one previously identified for the yeast Puf3 protein. Hence, the data obtained from the yeast and Drosophila studies provided an additional source for considering their evolution. For instance, conservation of amino acid residues in the RNA-binding domain (the PUM-homology domain) between homologous PUF proteins correlated with identified core motifs in 3'-UTR of mRNA targets. However, the proteins encoded by the mRNA targets appeared not to be particularly conserved. This discordance suggested that acquisition or loss of RBP binding motifs in UTRs of genes may provide a surprisingly fluid evolutionary mechanism to modify post-transcriptional regulatory connections.

Ribonomic studies have now been conducted for more than 30 specific RBPs. Each of the analyzed RBPs has a unique RNA binding spectrum comprised of 20–1000 distinct transcripts that often share functionally related themes. The spectra of targets overlap with other RBPs, suggesting combinatorial binding of RBPs. Occasionally, sequence or structural elements could be identified among mRNA targets using bioinformatics tools, and novel physiological consequences were discovered. The ribonomics approach has recently been implemented on the argonaute (Ago) protein family to discover novel mRNAs that potentially undergo miRNA dependent regulation. Although the number of detectable Ago-associated mRNAs was low (~90 messages) compared to the thousands of genes expected to undergo miRNA dependent regulation, the comparison of Ago-associated mRNAs in wild-type and miRNA mutants may provide a tool to decipher miRNA-specific targets. Besides specific RBPs, ribonomic approaches have also been applied to 'general' RNA-binding proteins for the identification of messages expressed in particular tissues or cell-types. Affinity-tagged poly(A) binding protein (PABP) was expressed with tissue-specific promoters to identify muscle- or ciliated sensory neuron-specific transcripts in the worm Caenorhabditis elegans , and mRNAs in photoreceptor cells of flies. The method was also used to measure gene expression of endothelial cells that were co-cultured with breast tumor cells. A similar approach with tagged ribosomal proteins may become another tool to determine gene expression in specific cells.

The application of genomic tools to study post-transcriptional gene regulation suggests additional levels of coordination and regulation that are beyond the traditional view of 'equally treated' cellular mRNAs that are similar processed, exported, and eventually translated in the cytoplasm. The decay, localization and translation of mRNA seem to undergo coordinate control by regulatory programs, which may be embedded in a multifaceted post-transcriptional regulatory system. The properties of this system are controlled by RNA-binding proteins or non-coding RNAs (e.g., microRNAs) that coordinate functionally related sets of mRNAs through binding to sequence elements in the RNA. Considering the hundreds of RBPs encoded in eukaryotic genomes, post-transcriptional control may be comparable in its richness and complexity to transcriptional regulatory systems. This provides a means to link RNA regulation to other cellular regulons such as signal transduction pathways allowing rapid and efficient reprogramming of gene expression in response to changing physiological conditions.

Further analysis of RBPs and their target RNAs may finally lead to a map of the proposed post-transcriptional regulatory system. However, besides the architecture, it will also be important to study the plasticity and dynamics of this regulation by measuring how it reacts in response to environmental or developmental changes, and how it is perturbed in certain diseases. Finally, a major challenge will be to connect the different levels of gene expression systems though large-scale data integration.

## Translation

The genes in DNA encode protein molecules, which are the "workhorses" of the cell, carrying out all the functions necessary for life. For example, enzymes, including those that metabolize nutrients and synthesize new cellular constituents, as well as DNA polymerases and other enzymes that make copies of DNA during cell division, are all proteins.

In the simplest sense, expressing a gene means manufacturing its corresponding protein, and this multi-layered process has two major steps. In the first step, the information in DNA is transferred to a messenger RNA (mRNA) molecule by way of a process called transcription. During transcription, the DNA of a gene serves as a template for complementary base-pairing, and an enzyme called RNA polymerase II catalyzes the formation of a pre-mRNA molecule, which is then processed to form mature mRNA. The resulting mRNA is a single-stranded copy of the gene, which next must be translated into a protein molecule.

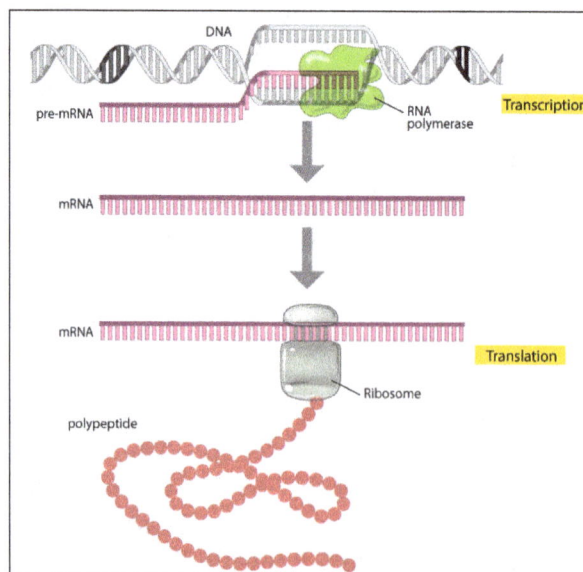

Figure: A gene is expressed through the processes of transcription and translation.

During transcription, the enzyme RNA polymerase (green) uses DNA as a template to produce a pre-mRNA transcript (pink). The pre-mRNA is processed to form a mature mRNA molecule that can be translated to build the protein molecule (polypeptide) encoded by the original gene.

During translation, which is the second major step in gene expression, the mRNA is "read" according to the genetic code, which relates the DNA sequence to the amino acid sequence in proteins. Each group of three bases in mRNA constitutes a codon, and each codon specifies a particular amino acid (hence, it is a triplet code). The mRNA sequence is thus used as a template to assemble—in order—the chain of amino acids that form a protein.

Figure: The amino acids specified by each mRNA codon. Multiple codons can code for the same amino acid.

The codons are written 5′ to 3′, as they appear in the mRNA. AUG is an initiation codon; UAA, UAG, and UGA are termination (stop) codons.

But where does translation take place within a cell? What individual substeps are a part of this process? And does translation differ between prokaryotes and eukaryotes? The answers to questions such as these reveal a great deal about the essential similarities between all species.

## Site of Translation

Within all cells, the translation machinery resides within a specialized organelle called the ribosome. In eukaryotes, mature mRNA molecules must leave the nucleus and travel to the cytoplasm, where the ribosomes are located. On the other hand, in prokaryotic organisms, ribosomes can attach to mRNA while it is still being transcribed. In this situation, translation begins at the 5′ end of the mRNA while the 3′ end is still attached to DNA.

In all types of cells, the ribosome is composed of two subunits: the large (50S) subunit and the small (30S) subunit (S, for svedberg unit, is a measure of sedimentation velocity and, therefore, mass). Each subunit exists separately in the cytoplasm, but the two join together on the mRNA molecule. The ribosomal subunits contain proteins and specialized RNA molecules—specifically, ribosomal RNA (rRNA) and transfer RNA (tRNA). The tRNA molecules are adaptor molecules—they have one end that can read the triplet code in the mRNA through complementary base-pairing, and another end that attaches to a specific amino acid. The idea that tRNA was an adaptor molecule was first proposed by Francis Crick, co-discoverer of DNA structure, who did much of the key work in deciphering the genetic code.

Within the ribosome, the mRNA and aminoacyl-tRNA complexes are held together closely, which facilitates base-pairing. The rRNA catalyzes the attachment of each new amino acid to the growing chain.

## Beginning of mRNA is not Translated

Interestingly, not all regions of an mRNA molecule correspond to particular amino acids. In particular, there is an area near the 5′ end of the molecule that is known as the untranslated region (UTR) or leader sequence. This portion of mRNA is located between the first nucleotide that is transcribed and the start codon (AUG) of the coding region, and it does not affect the sequence of amino acids in a protein.

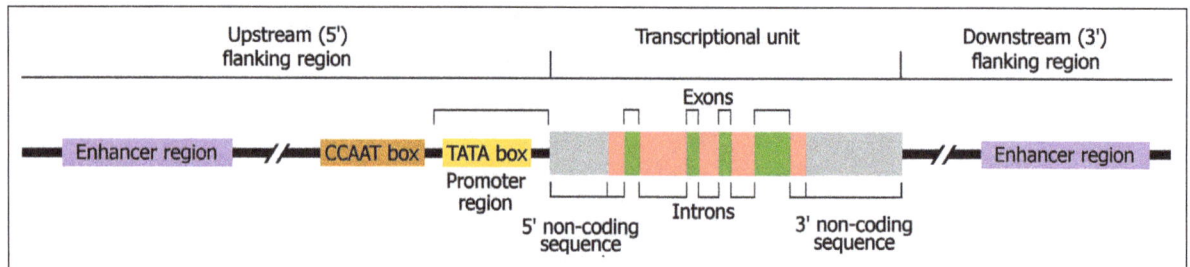

Figure: A DNA transcription unit.

So, what is the purpose of the UTR? It turns out that the leader sequence is important because it contains a ribosome-binding site. In bacteria, this site is known as the Shine-Dalgarno box (AG-GAGG), after scientists John Shine and Lynn Dalgarno, who first characterized it. A similar site in vertebrates was characterized by Marilyn Kozak and is thus known as the Kozak box. In bacterial mRNA, the 5′ UTR is normally short; in human mRNA, the median length of the 5′ UTR is about 170 nucleotides. If the leader is long, it may contain regulatory sequences, including binding sites for proteins, that can affect the stability of the mRNA or the efficiency of its translation.

A DNA transcription unit is composed, from its 3′ to 5′ end, of an RNA-coding region (pink rectangle) flanked by a promoter region (green rectangle) and a terminator region (black rectangle). Regions to the left, or moving towards the 3′ end, of the transcription start site are considered upstream; regions to the right, or moving towards the 5′ end, of the transcription start site are considered downstream.

## Translation Begins after the Assembly of a Complex Structure

The translation of mRNA begins with the formation of a complex on the mRNA. First, three initiation factor proteins (known as IF1, IF2, and IF3) bind to the small subunit of the ribosome. This preinitiation complex and a methionine-carrying tRNA then bind to the mRNA, near the AUG start codon, forming the initiation complex.

When translation begins, the small subunit of the ribosome and an initiator tRNA molecule assemble on the mRNA transcript. The small subunit of the ribosome has three binding sites: an amino acid site (A), a polypeptide site (P), and an exit site (E). The initiator tRNA molecule carrying the amino acid methionine binds to the AUG start codon of the mRNA transcript at the ribosome's P site where it will become the first amino acid incorporated into the growing polypeptide chain.

Here, the initiator tRNA molecule is shown binding after the small ribosomal subunit has assembled on the mRNA; the order in which this occurs is unique to prokaryotic cells. In eukaryotes, the free initiator tRNA first binds the small ribosomal subunit to form a complex. The complex then binds the mRNA transcript, so that the tRNA and the small ribosomal subunit bind the mRNA simultaneously.

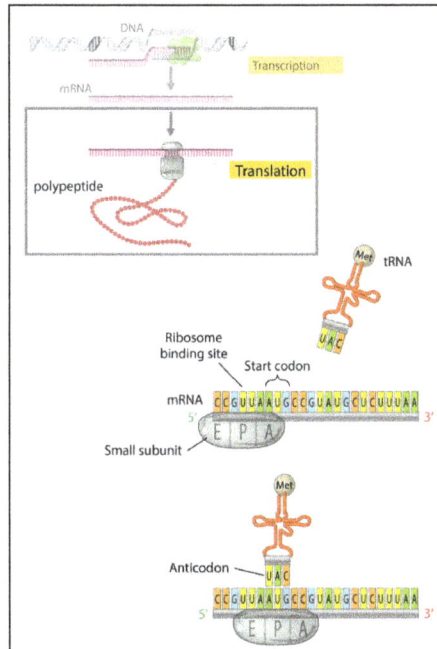

Figure: The translation initiation complex.

Although methionine (Met) is the first amino acid incorporated into any new protein, it is not always the first amino acid in mature proteins—in many proteins, methionine is removed after translation. In fact, if a large number of proteins are sequenced and compared with their known gene sequences, methionine (or for my methionine) occurs at the N-terminus of all of them. However, not all amino acids are equally likely to occur second in the chain, and the second amino acid influences whether the initial methionine is enzymatically removed. For example, many proteins begin with methionine followed by alanine. In both prokaryotes and eukaryotes, these proteins have the methionine removed, so that alanine becomes the N-terminal amino acid. However, if the second amino acid is lysine, which is also frequently the case, methionine is not removed (at least in the sample proteins that have been studied thus far). These proteins therefore begin with methionine followed by lysine.

Table shows the N-terminal sequences of proteins in prokaryotes and eukaryotes, based on a sample of 170 prokaryotic and 120 eukaryotic proteins. In the table, M represents methionine, A represents alanine, K represents lysine, S represents serine, and T represents threonine.

Table: N-Terminal Sequences of Proteins.

| N-Terminal Sequence | Percent of Prokaryotic Proteins with This Sequence | Percent of Eukaryotic Proteins with This Sequence |
|---|---|---|
| MA* | 28.24% | 19.17% |
| MK** | 10.59% | 2.50% |

| MS* | 9.41% | 11.67% |
|---|---|---|
| MT* | 7.65% | 6.67% |
| * Methionine was removed in all of these proteins  ** Methionine was not removed from any of these proteins | | |

Once the initiation complex is formed on the mRNA, the large ribosomal subunit binds to this complex, which causes the release of IFs (initiation factors). The large subunit of the ribosome has three sites at which tRNA molecules can bind. The A (amino acid) site is the location at which the aminoacyl-tRNA anticodon base pairs up with the mRNA codon, ensuring that correct amino acid is added to the growing polypeptide chain. The P (polypeptide) site is the location at which the amino acid is transferred from its tRNA to the growing polypeptide chain. Finally, the E (exit) site is the location at which the "empty" tRNA sits before being released back into the cytoplasm to bind another amino acid and repeat the process. The initiator methionine tRNA is the only aminoacyl-tRNA that can bind in the P site of the ribosome, and the A site is aligned with the second mRNA codon. The ribosome is thus ready to bind the second aminoacyl-tRNA at the A site, which will be joined to the initiator methionine by the first peptide bond.

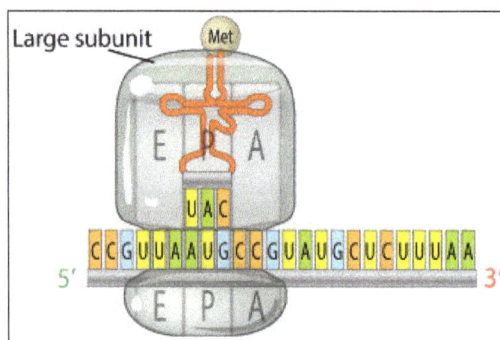

Figure: The large ribosomal subunit binds to the small ribosomal subunit to complete the initiation complex.

The initiator tRNA molecule, carrying the methionine amino acid that will serve as the first amino acid of the polypeptide chain, is bound to the P site on the ribosome. The A site is aligned with the next codon, which will be bound by the anticodon of the next incoming tRNA.

## Elongation Phase

The next phase in translation is known as the elongation phase. First, the ribosome moves along the mRNA in the 5′-to-3′direction, which requires the elongation factor G, in a process called translocation. The tRNA that corresponds to the second codon can then bind to the A site, a step that requires elongation factors (in E. coli, these are called EF-Tu and EF-Ts), as well as guanosine triphosphate (GTP) as an energy source for the process. Upon binding of the tRNA-amino acid complex in the A site, GTP is cleaved to form guanosine diphosphate (GDP), then released along with EF-Tu to be recycled by EF-Ts for the next round.

Next, peptide bonds between the now-adjacent first and second amino acids are formed through a peptidyl transferase activity. For many years, it was thought that an enzyme catalyzed this step, but recent evidence indicates that the transferase activity is a catalytic function of rRNA. After the peptide bond is formed, the ribosome shifts, or translocates, again, thus causing the tRNA to occupy the E site. The tRNA is then released to the cytoplasm to pick up another amino acid. In addition, the A site is now empty and ready to receive the tRNA for the next codon.

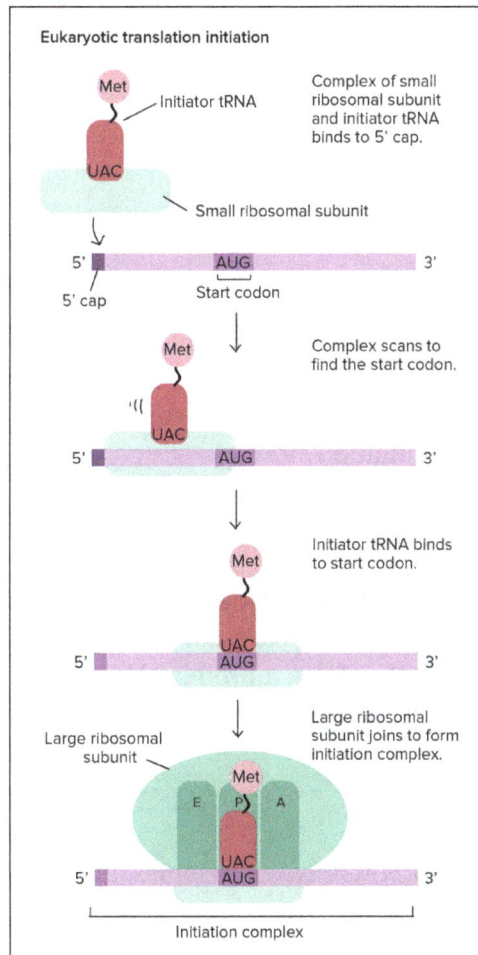

Eukaryotic translation initiation

This process is repeated until all the codons in the mRNA have been read by tRNA molecules, and the amino acids attached to the tRNAs have been linked together in the growing polypeptide chain in the appropriate order. At this point, translation must be terminated, and the nascent protein must be released from the mRNA and ribosome.

## Termination of Translation

There are three termination codons that are employed at the end of a protein-coding sequence in mRNA: UAA, UAG, and UGA. No tRNAs recognize these codons. Thus, in the place of these tRNAs, one of several proteins, called release factors, binds and facilitates release of the mRNA from the ribosome and subsequent dissociation of the ribosome.

## Comparing Eukaryotic and Prokaryotic Translation

The translation process is very similar in prokaryotes and eukaryotes. Although different elongation, initiation, and termination factors are used, the genetic code is generally identical. As previously noted, in bacteria, transcription and translation take place simultaneously, and mRNAs are relatively short-lived. In eukaryotes, however, mRNAs have highly variable half-lives, are subject to modifications, and must exit the nucleus to be translated; these multiple steps offer additional opportunities to regulate levels of protein production, and thereby fine-tune gene expression.

# Prokaryotic Translation

## Site

Translation occurs in the cytoplasm where the ribosomes are located. Ribosomes are made of a small and large subunit which surrounds the mRNA. In prokaryotic translation 70S ribosomes with 30S and 50S subunits are used. The mRNA is synthesized from DNA only. In prokaryotes, there are several initiation and termination sites.

## Template

In translation, messenger RNA (mRNA) is decoded to produce a specific polypeptide according to the rules specified by the genetic code. This uses an mRNA sequence as a template to guide the synthesis of a chain of amino acids that form a protein. Many types of transcribed RNA, such as transfer RNA, ribosomal RNA, and small nuclear RNA are not necessarily translated into an amino acid sequence.

## Requirements

The translation process requires mRNA, rRNA, ribosomes, 20 kinds of amino acids and their specific tRNAs.

## Factors Involved

In prokaryotes, three factors are involved in the initiation of translation [IF 1, IF 2 and IF 3], one factor in the elongation of polypeptide chain and three factors in chain termination [RF1, RF2 and RF3].

## Enzymes Involved

Two types of enzymes are used in translation. Aminoacyl tRNA synthetase (an enzyme) catalyzes the bonding between specific tRNAs and the amino acids. The enzyme peptidyl transferase connects A site and P site by forming a peptide bond [the nitrogen carbon bond] during elongation phase.

## Codons Involved

In the process of translation two types of codons, viz., start codon and stop codons are involved. The codon, AUG, initiates the process of translation and one of three stop codons i.e. UAA, UAG or UGA is used for chain termination.

## Starting Amino Acid

In prokaryotes, starting amino acid is N-formyl methionine. Moreover, there is overlapping of transcription and translation.

## Mechanism of Translation in Prokaryotes

Translation process consists of three major phases or stages:

## Initiation

Initiation of translation in prokaryotes involves the assembly of the components of the translation system which are: the two ribosomal subunits (small and large), the mRNA to be translated, the first (formyl) aminoacyl tRNA (the tRNA charged with the first amino acid), GTP (as a source of energy), and three initiation factors (IF 1, IF 2 and IF 3) which help the assembly of the initiation complex.

The ribosome consists of three sites, the A site, the P site, and the E site. The A site is the point of entry for the aminoacyl tRNA (except for the first aminoacyl tRNA, fMet-tRNA$_f^{Met}$, which enters at the P site). The P site is where the peptidyl tRNA is formed in the ribosome. And the E site which is the exit site of the now uncharged tRNA after it gives its amino acid to the growing peptide chain.

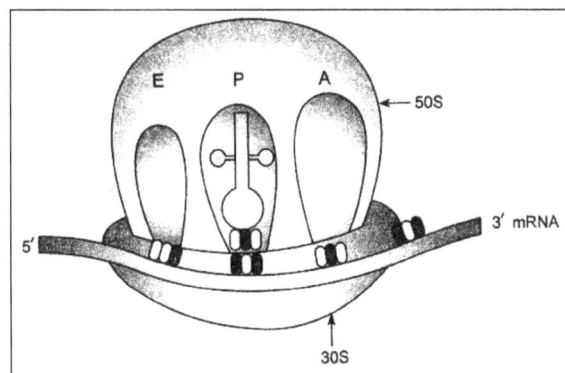

Figure: Initiation

Translation begins with the binding of the small ribosomal subunit to a specific sequence on the mRNA chain. Initiation of translation begins with the 50S and 30S ribosomal subunits. IF1 (initiation factor 1) blocks the A site to ensure that the IMet-tRNA can bind only to the P site and that no other aminoacyl-tRNA can bind in the A site during initiation, while IF3 blocks the E site and prevents the two subunits from associating.

IF2 is a small GTPase which binds fmet-tRNA$_f^{Met}$ and helps its binding with the small ribosomal subunit. The 3' end of the 16S rRNA of the small 30S ribosomal subunit recognizes the ribosomal binding site on the mRNA (Shine-Dalgarno sequence or SD), through its anti-SD sequence, 5-10 base pairs upstream of the start codon. The Shine-Dalgarno sequence is found only in prokaryotes.

This helps to correctly position the ribosome onto the mRNA so that the P site is directly on the AUG initiation codon. IF3 helps to position fMet-tRNA$_f^{met}$ into the P site, such that fMet-tRNA$_f^{met}$ interacts via base pairing with the mRNA initiation codon (AUG). Initiation ends as the large ribosomal subunit joins the complex causing the dissociation of initiation factors.

The small subunit binds via complementary base pairing between one of its internal subunits and the ribosome binding site. This site a sequence of about ten nucleotides on the mRNA. It is located anywhere from 5 and 11 nucleotides from the initiating codon [AUG],

After binding of the small subunit, a special tRNA molecule, called N-formyl methionine, or fMet, recognizes and binds to the initiator codon. Then the large subunit binds resulting in the formation of the initiation complex. As soon as the initiation complex is formed, the fMet-tRNA occupies the P site of the ribosome and the A site is left empty.

This entire initiation process is facilitated by extra proteins, called initiation factors that help with the binding of ribosomal subunits and tRNA to the mRNA chain.

## Elongation

This is the second phase or middle phase of translation. Elongation begins after the formation of the initiation complex. Elongation of the polypeptide chain involves addition of amino acids to the carboxyl end of the growing chain. The growing protein exits the ribosome through the polypeptide exit tunnel in the large subunit.

Elongation starts when the fmet-tRNA enters the P site, causing a conformational change which opens the A site for the new aminoacyl-tRNA to bind. This binding is facilitated by elongation factor-T4 (EF-T4), a small GTPase. Now the P site contains the beginning of the peptide chain of the protein to be encoded and the A site has the next amino acid to be added to the peptide chain. The growing polypeptide connected to the tRNA in the P site is detached from the tRNA in the P site and a peptide bond is formed between the last amino acids of the polypeptide and the amino acid still attached to the tRNA in the A site.

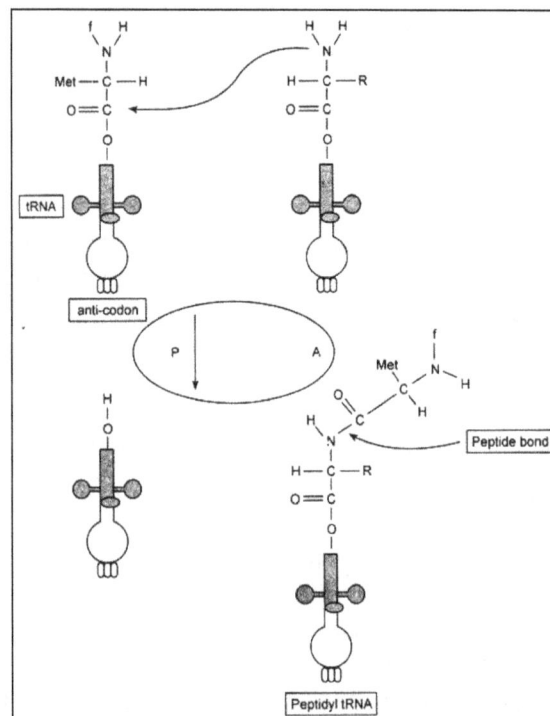

Figure: Peptide Formation.

This process, known as peptide bond formation, is catalyzed by a ribozyme, peptidyltransferase, an activity intrinsic to the 23S ribosomal RNA in the 50S ribosomal subunit. Now, the A site has newly formed peptide, while the P site has an unloaded tRNA (tRNA with no amino acids). In the final stage of elongation, translocation, the ribosome moves 3 nucleotides towards the 3' end of mRNA.

Since tRNAs are linked to mRNA by codon-anticodon base-pairing, tRNAs move relative to the ribosome taking the nascent polypeptide from the A site to the P site and moving the uncharged tRNA to the E exit site. This process is catalyzed by elongation factor G (EF-G). The ribosome continues to translate the remaining codons on the mRNA as more aminoacyl-tRNA binds to the A site, until the ribosome reaches a stop codon on mRNA(UAA, UGA, or UAG).

When the A site opens again, the next appropriate aminoacyl tRNA can bind there and the same reaction takes place, yielding a three-amino acid peptide chain. This process repeats, creating a polypeptide chain in the P site of the ribosome. A single ribosome can translate 60 nucleotides per second. This speed can be vastly augmented when ribosomes unite together to form polyribosomes.

## Termination

This is the last phase of translation. Termination occurs when one of the three termination codons moves into the A site. These codons are not recognized by any tRNAs. Instead, they are recognized by proteins called release factors, namely RF1 (recognizing the UAA and UAG stop codons) or RF2 (recognizing the UAA and UGA stop codons).

These factors trigger the hydrolysis of the ester bond in peptidyl-tRNA and the release of the newly synthesized protein from the ribosome. A third release factor RF-3 catalyzes the release of RF-1 and RF-2 at the end of the termination process.

# Eukaryotic Translation

## Site

Translation occurs in the cytoplasm where the ribosomes are located. Ribosomes are made of a small and large subunit which surrounds the mRNA. In eukaryotic translation 80S ribosomes with 40S and 60S subunits are used. The mRNA is synthesized from DNA only. In eukaryotes, there is single initiation and termination site.

## Template

This uses an mRNA sequence as a template to guide the synthesis of a chain of amino acids that form a protein. Many types of transcribed RNA, such as transfer RNA, ribosomal RNA, and small nuclear RNA are not necessarily translated into an amino acid sequence.

## Requirements

The translation process requires mRNA, rRNA, ribosomes, 20 kinds of amino acids and their specific tRNAs.

## Factors Involved

In eukaryotes, several factors are used in chain initiation such as eIF2, eIF3, eIF4A, eIF4E, eIF4F and elF 4G. Two factors [EF-1 and EF-2] are used in chain elongation. There is a single release factor RF for recognition of three termination codons [UAA, UAG and UGA].

## Enzymes Involved

In eukaryotes, two types of enzymes are used in translation. Aminoacyl tRNA synthetase (an enzyme) catalyzes the bonding between specific tRNAs and the amino acids. The enzyme peptidyl transferase connects. A site and P site by forming a peptide bond [the nitrogen carbon bond] during elongation phase.

## Codons Involved

In the process of translation two types of codons, viz., start codorl and stop codons are involved. The codon, AUG. initiates the process of translation and one of three stop codons i.e. UAA, UAG, or UGA is used for chain termination. In eukaryotes and archaea, the amino acid encoded by the start codon is methionine.

## Starting Amino Acid

In eukaryotes, starting amino acid is methionine. Moreover, there is no overlapping of transcription and translation.

## Mechanism of Translation in Eukaryotes

The mechanism of translation in eukaryotes is similar to that of prokaryotes in several aspects. Translation process consists of three phases or stages.

Initiation: The process of initiation of translation in eukaryotes is of two types.

1. Cap-dependent initiation, and

2. Cap-independent initiation.

## Cap-dependent Initiation

Initiation of translation usually involves the interaction of certain key proteins with a special tag bound to the 5′-end of an mRNA molecule, the 5′ cap. The protein factors bind the small ribosomal subunit (also referred to as the 40S subunit), and these initiation factors hold the mRNA in place.

The eukaryotic Initiation Factor 3 (eIF3) is associated with the small ribosomal subunit, and plays a role in keeping the large ribosomal subunit from prematurely binding. The factor eIF3 also interacts with the eIF4F complex which consists of three other initiation factors [eIF4A, eIF4E and eIF4G]. The factor eIF4G is a protein which directly associates with both eIF3 and the other two components.

The eIF4E is the cap-binding protein. It is the rate-limiting step of capdependent initiation, and is often cleaved from the complex by some viral proteases to limit the cell's ability to translate its own transcripts. The eIF4A is an ATP-dependent RNA helicase, which aids the ribosome in resolving certain secondary structures formed by the mRNA transcript. There is another protein associated with the eIF4F complex called the Poly-A Binding Protein (PABP), which binds the poly-A tail of most eukaryotic mRNA molecules. This protein is considered to play a role in circularization of the mRNA during translation.

This pre-initiation complex (43S subunit, or the 40S and mRNA) along with protein factors move along the mRNA chain towards its 3'-end. It scans for the 'start' codon (typically AUG) on the mRNA. The start codon indicates the site where the mRNA will begin coding for the protein. In eukaryotes and archaea, the amino acid encoded by the start codon is methionine. The initiator tRNA charged with Met forms pan of the ribosomal complex and thus all proteins start with this amino acid. The Met-charged initiator tRNA is brought to the P-site of the small ribosomal subunit by eukaryotic Initiation Factor 2 (eIF2). It hydrolyzes GTP, and signals for the dissociation of several factors from the small ribosomal subunit which results in the association of the large subunit (or the 60S subunit). The complete ribosome (80S) then commences translation elongation, during which the sequence between the 'start' and 'stop' codons is translated from mRNA into an amino acid sequence. In this way a protein is synthesized.

## Cap-independent Initiation

This is lesser known method of translation in eukaryotes. This method of translation has been recently discovered. It has been found to be important in conditions that require the translation of specific mRNAs. It works despite cellular stress or the inability to translate most mRNAs. Examples of such type of translation are factors responding to apoptosis and stress-induced responses. The best studied example of the cap-independent mode of translation initiation in eukaryotes is the Internal Ribosome Entry Site (IRES) approach. The main difference between cap-independent translation and cap-dependent translation is that the former does not require the ribosome to start scanning from the 5' end of the mRNA cap until the start codon. The ribosome can be trafficked to the start site by ITAFs (IRES trans-acting factors) bypassing the need to scan from the 5' end of the un-translated region of the mRNA.

## Elongation

Elongation is dependent on eukaryotic elongation factors At the end of the initiation step, the mRNA is positioned so that the next codon can be translated during the elongation stage of protein synthesis. The initiator tRNA occupies the P site in the ribosome; and the A site is ready to receive an aminoacyl-tRNA. During chain elongation, each additional amino acid is added to the nascent polypeptide chain in a three-step micro-cycle.

The steps in this micro-cycle are:

1. Positioning the correct aminoacyl-tRNA in the A site of the ribosome;

2. Forming the peptide bond;

3. Shifting the mRNA by one codon relative to the ribosome.

The translation machinery works relatively slowly compared to the enzyme systems that catalyze

DNA replication. Proteins are synthesised at a rate of only 18 amino acid residues per second, whereas bacterial replisomes synthesize DNA at a rate of 1,000 nucleotides per second.

This difference in rate reflects, in part, the difference between polymerizing four types of nucleotides to make nucleic acids and polymerizing 20 types of amino acids to make proteins. Testing and rejecting incorrect aminoacyl- tRNA molecules takes time and slows protein synthesis. The rate of transcription in prokaryotes is approximately 55 nucleotides per second, which corresponds to about 18 codons per second, or the same rate at which the mRNA is translated.

In bacteria, translation initiation occurs as soon as the 5′ end of an mRNA is synthesized, and translation and transcription are coupled. This tight coupling is not possible in eukaryotes because transcription and translation are carried out in separate compartments of the cell (the nucleus and cytoplasm). Eukaryotic mRNA precursors must be processed in the nucleus (e.g., capping, polyadenylation, splicing) before they are exported to the cytoplasm for translation.

## Termination

This is the last phase of translation. Termination occurs when one of the three termination codons moves into the A site. These codons are not recognized by any tRNAs.

Termination of elongation is dependent on eukaryotic release factors. In eukaryotes, there is only one release factor that is eRF, which recognizes all three stop codons [in place of RF1, RF2, or RF3 factors in prokaryotes]. However, the overall process of termination is similar to that of prokaryotes.

## Prokaryotic versus Eukaryotic Translation

The basic steps involved in protein synthesis are similar in both prokaryotes and eukaryotes. However, protein synthesis differs in several aspects in these two groups.

| S.No. | Particulars | Prokaryotes | Eukaryotes |
|-------|-------------|-------------|------------|
| 1. | Polymerase used | One | Several |
| 2. | Ribosomes used | 70S with 30S and 50S subunits | 80S with 40S and 60S subunits |
| 3. | Transcription and translation | Can overlap | Do not overlap |
| 4. | Starting amino acid | N-formyl methionine | Methioine |
| 5. | Initiation and termination sites | Several | Single |
| 6. | Initiation factors used | IF 1, IF 2, and IF 3 | eLF2, eIF3, elF 4A, elF 4E elF 4F and elF 4 |
| 7. | Chain elongation factors | EF-T4, EF-TS and EF_G | EF1 and EF2 |
| 8. | Release Factors | RF 1, RF 2 and RF 3 | RF |
| 9. | Synthesis of mRNA | From DNA ir RNA | From RNA only |

## References

- Gene-expression-14121669, scitable: nature.com, Retrieved 2 March, 2019

- Gene-expression-is-analyzed-by-tracking-rna-6525038, scitable: nature.com, Retrieved 18 May, 2019

- Outcome-prokaryotic-gene-regulation, biology: courses.lumenlearning.com, Retrieved 20 July, 2019

- Rna-processing-in-eukaryotes, boundless-biology: courses.lumenlearning.com, Retrieved 22 January, 2019

- Translation-dna-to-mrna-to-protein-393, scitable: nature.com, Retrieved 26 March, 2019

- Translation-in-prokaryotes-genetics, 38022, prokaryotes, cell: biologydiscussion.com, Retrieved 6 June, 2019

- Translation-in-eukaryotes-genetics/37991, eukaryotes, organisms: biologydiscussion.com, Retrieved 14 August, 2019s

# Chapter 6

## DNA Editing

There are a number of different techniques and processes which are involved in the editing or modification of the DNA. A few of them are DNA mutations, DNA repair, genetic recombination, RNA interference and CRISPR. This chapter closely examines the key concepts related to these techniques to provide an extensive understanding of the subject.

## DNA Mutations

Since mutations are simply changes in DNA, in order to understand how mutations work, you need to understand how DNA does its job. Your DNA contains a set of instructions for "building" a human. These instructions are inscribed in the structure of the DNA molecule through a genetic code. It works like this:

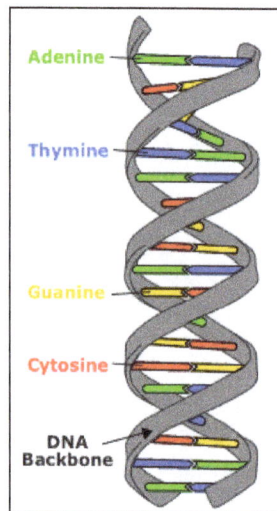

DNA is made of a long sequence of smaller units strung together. There are four basic types of unit: A, T, G, and C. these letters represents the type of base each unit carries: adenine, thymine, guanine, and cytosine. The sequence of these bases encodes instructions. Some parts of your DNA are control centers for turning genes on and off, some parts have no function, and some parts have a function that we don't understand yet. Other parts of your DNA are genes that carry the instructions for making proteins — which are long chains of amino acids. These proteins help build an organism.

Protein-coding DNA can be divided into codons — sets of three bases that specify an amino acid or signal the end of the protein. Codons are identified by the bases that make them up — in the example at right, GCA, for guanine, cytosine, and adenine. The cellular machinery uses these instructions to assemble a string of corresponding amino acids (one amino acid for each three bases) that form a protein. The amino acid that corresponds to "GCA" is called alanine; there are twenty different amino acids synthesized this way in humans. "Stop" codons signify the end of the newly built protein.

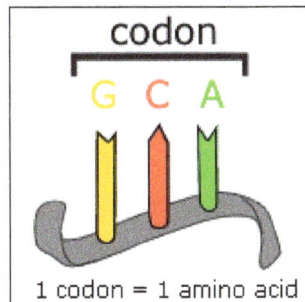

After the protein is built based on the sequence of bases in the gene, the completed protein is released to do its job in the cell.

## Types of Mutations

There are many different ways that DNA can be changed, resulting in different types of mutation.

### Substitution

A substitution is a mutation that exchanges one base for another (i.e., a change in a single "chemical letter" such as switching an A to a G). Such a substitution could:

- Change a codon to one that encodes a different amino acid and cause a small change in the protein produced. For example, sickle cell anemia is caused by a substitution in the beta-hemoglobin gene, which alters a single amino acid in the protein produced.

- Change a codon to one that encodes the same amino acid and causes no change in the protein produced. These are called silent mutations.

- Change an amino-acid-coding codon to a single "stop" codon and cause an incomplete protein. This can have serious effects since the incomplete protein probably won't function.

### Insertion

Insertions are mutations in which extra base pairs are inserted into a new place in the DNA.

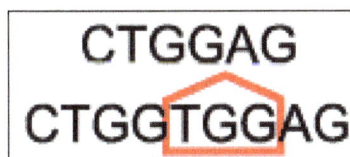

## Deletion

Deletions are mutations in which a section of DNA is lost, or deleted.

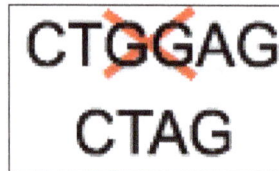

CTGGAG

CTAG

## Frameshift

The fat cat sat

hef atc ats at

Since protein-coding DNA is divided into codons three bases long, insertions and deletions can alter a gene so that its message is no longer correctly parsed. These changes are called frame shifts. For example, consider the sentence, "The fat cat sat." Each word represents a codon. If we delete the first letter and parse the sentence in the same way, it doesn't make sense.

In frameshifts, a similar error occurs at the DNA level, causing the codons to be parsed incorrectly. This usually generates truncated proteins that are as useless as "hef atc ats at" is uninformative.

## Causes of Mutations

Mutations happen for several reasons.

### DNA Fails to Copy Accurately

Most of the mutations that we think matter to evolution are "naturally-occurring." For example, when a cell divides, it makes a copy of its DNA — and sometimes the copy is not quite perfect. That small difference from the original DNA sequence is a mutation.

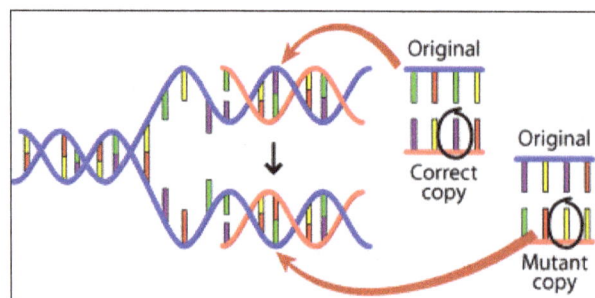

### External Influences can Create Mutations

Mutations can also be caused by exposure to specific chemicals or radiation. These agents cause the DNA to break down. This is not necessarily unnatural — even in the most isolated and pristine environments, DNA breaks down. Nevertheless, when the cell repairs the DNA, it might not do a perfect job of the repair. So the cell would end up with DNA slightly different than the original DNA and hence, a mutation.

## Effects of Mutations

An apple with a somatic mutation. Since all cells in our body contain DNA, there are lots of places for mutations to occur; however, some mutations cannot be passed on to offspring and do not matter for evolution. Somatic mutations occur in non-reproductive cells and won't be passed onto offspring. For example, the golden color on half of this Red Delicious apple was caused by a somatic mutation. Its seeds will not carry the mutation. The only mutations that matter to large-scale evolution are those that can be passed on to offspring. These occur in reproductive cells like eggs and sperm and are called germ line mutations.

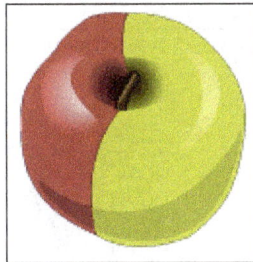

## Effects of Germ Line Mutations

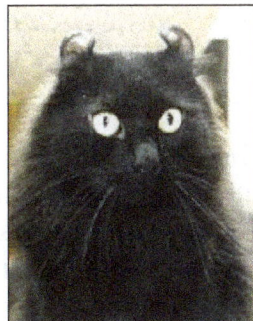

A single germ line mutation can have a range of effects:

1. No change occurs in phenotype: Some mutations don't have any noticeable effect on the phenotype of an organism. This can happen in many situations: perhaps the mutation

occurs in a stretch of DNA with no function, or perhaps the mutation occurs in a protein-coding region, but ends up not affecting the amino acid sequence of the protein.

2.  Small change occurs in phenotype: A single mutation caused this cat's ears to curl backwards slightly.

3.  Big change occurs in phenotype: Some really important phenotypic changes, like DDT resistance in insects are sometimes caused by single mutations. A single mutation can also have strong negative effects for the organism. Mutations that cause the death of an organism are called lethals — and it doesn't get more negative than that.

## Little Mutations with Big Effects: Mutations to Control Genes

Mutations are often the victims of bad press — unfairly stereotyped as unimportant or as a cause of genetic disease. While many mutations do indeed have small or negative effects, another sort of mutation gets less airtime. Mutations to control genes can have major (and sometimes positive) effects. Some regions of DNA control other genes, determining when and where other genes are turned "on". Mutations in these parts of the genome can substantially change the way the organism is built. The difference between a mutation to a control gene and a mutation to a less powerful gene is a bit like the difference between whispering an instruction to the trumpet player in an orchestra versus whispering it to the orchestra's conductor. The impact of changing the conductor's behavior is much bigger and more coordinated than changing the behavior of an individual orchestra member. Similarly, a mutation in a gene "conductor" can cause a cascade of effects in the behavior of genes under its control.

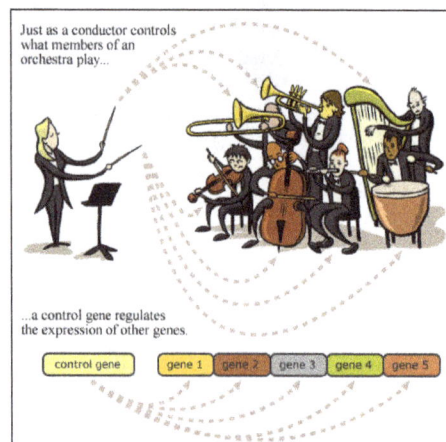

Many organisms have powerful control genes that determine how the body is laid out. For example, Hox genes are found in many animals (including flies and humans) and designate where the head goes and which regions of the body grow appendages. Such masters controls genes help direct the building of body "units," such as segments, limbs, and eyes. So evolving a major change in basic body layout may not be so unlikely; it may simply require a change in a Hox gene and the favor of natural selection.

## Case Study of the Effects of Mutation: Sickle Cell Anemia

Sickle cell anemia is a genetic disease with severe symptoms, including pain and anemia. The disease is caused by a mutated version of the gene that helps make hemoglobin — a protein that carries oxygen in red blood cells. People with two copies of the sickle cell gene have the disease. People who carry only one copy of the sickle cell gene do not have the disease, but may pass the gene on to their children. The mutations that cause sickle cell anemia have been extensively studied and demonstrate how the effects of mutations can be traced from the DNA level up to the level of the whole organism. Consider someone carrying only one copy of the gene. She does not have the disease, but the gene that she carries still affects her, her cells, and her proteins.

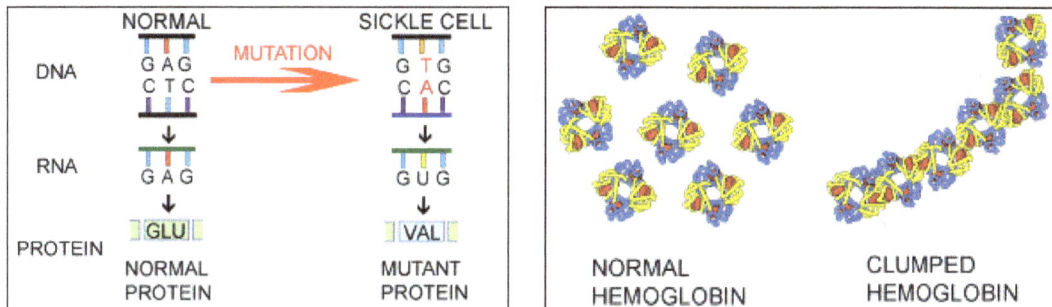

There are effects at the DNA level.                    There are effects at the protein level.

Normal hemoglobin (left) and hemoglobin in sickled red blood cells (right) look different; the mutation in the DNA slightly changes the shape of the hemoglobin molecule, allowing it to clump together.

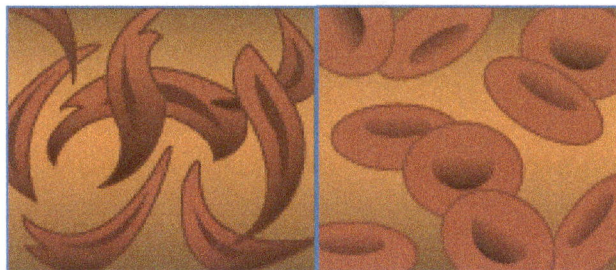

Normal red blood cells (right) and sickle cells (left)

1.  There are effects at the cellular level: When red blood cells carrying mutant hemoglobin are deprived of oxygen, they become "sickle-shaped" instead of the usual round shape. This shape can sometimes interrupt blood flow.

2.  There are negative effects at the whole organism level: Under conditions such as high elevation and intense exercise, a carrier of the sickle cell allele may occasionally show symptoms such as pain and fatigue.

3.  There are positive effects at the whole organism level: Carriers of the sickle cell allele are resistant to malaria, because the parasites that cause this disease are killed inside sickle-shaped blood cells.

This is a chain of causation. What happens at the DNA level propagates up to the level of the complete organism. This example illustrates how a single mutation can have a large effect, in this case, both a positive and a negative one. But in many cases, evolutionary change is based on the accumulation of many mutations, each having a small effect. Whether the mutations are large or small, however, the same chain of causation applies: changes at the DNA level propagate up to the phenotype.

## Mutations are Random

Mutations can be beneficial, neutral, or harmful for the organism, but mutations do not "try" to supply what the organism "needs." Factors in the environment may influence the rate of mutation but are not generally thought to influence the direction of mutation. For example, exposure to harmful chemicals may increase the mutation rate, but will not cause more mutations that make the organism resistant to those chemicals. In this respect, mutations are random — whether a particular mutation happens or not is unrelated to how useful that mutation would be.

For example, in the U.S. where people have access to shampoos with chemicals that kill lice, we have a lot of lice that are resistant to those chemicals. There are two possible explanations for this:

### Hypothesis A

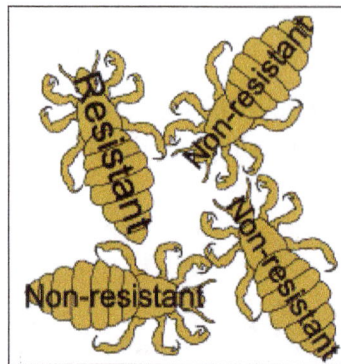

Resistant strains of lice were always there and are just more frequent now
because all the non-resistant lice died a sudsy death.

### Hypothesis B

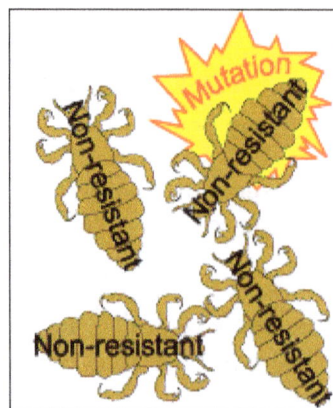

Exposure to lice shampoo actually caused mutations for resistance to the shampoo.

Scientists generally think that the first explanation is the right one and that directed mutations, the second possible explanation relying on non-random mutation, is not correct. Researchers have performed many experiments in this area. Though results can be interpreted in several ways, none unambiguously support directed mutation. Nevertheless, scientists are still doing research that provides evidence relevant to this issue.

In addition, experiments have made it clear that many mutations are in fact random, and did not occur because the organism was placed in a situation where the mutation would be useful. For example, if you expose bacteria to an antibiotic, you will likely observe an increased prevalence of antibiotic resistance. Esther and Joshua Lederberg determined that many of these mutations for antibiotic resistance existed in the population even before the population was exposed to the antibiotic — and that exposure to the antibiotic did not cause those new resistant mutants to appear.

## Lederberg Experiment

In 1952, Esther and Joshua Lederberg performed an experiment that helped show that many mutations are random, not directed. In this experiment, they capitalized on the ease with which bacteria can be grown and maintained. Bacteria grow into isolated colonies on plates. These colonies can be reproduced from an original plate to new plates by "stamping" the original plate with a cloth and then stamping empty plates with the same cloth. Bacteria from each colony are picked up on the cloth and then deposited on the new plates by the cloth.

Esther and Joshua hypothesized that antibiotic resistant strains of bacteria surviving an application of antibiotics had the resistance before their exposure to the antibiotics, not as a result of the exposure. Their experimental set-up is summarized below:

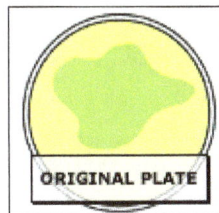

1. Bacteria are spread out on a plate, called the "original plate."

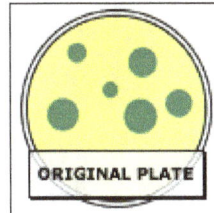

2. They are allowed to grow into several different colonies.

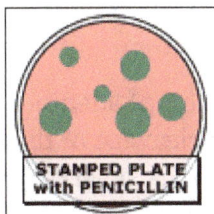

3. This layout of colonies is stamped from the original plate onto a new plate that contains the antibiotic penicillin.

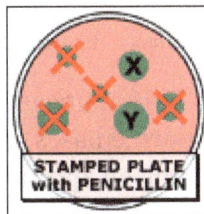

4. Colonies X and Y on the stamped plate survive. They must carry a mutation for penicillin resistance.

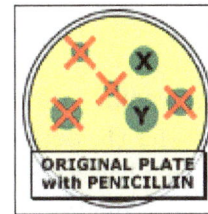

5. The Lederbergs set out to answer the question, "did the colonies on the new plate evolve antibiotic resistance because they were exposed to penicillin?"

When the original plate is washed with penicillin, the same colonies (those in position X and Y) live — even though these colonies on the original plate have never encountered penicillin before. So the penicillin-resistant bacteria were there in the population before they encountered penicillin. They did not evolve resistance in response to exposure to the antibiotic.

# DNA Repair

DNA, like any other molecule, can undergo a variety of chemical reactions. Because DNA uniquely serves as a permanent copy of the cell genome, however, changes in its structure are of much greater consequence than are alterations in other cell components, such as RNAs or proteins. Mutations can result from the incorporation of incorrect bases during DNA replication. In addition, various chemical changes occur in DNA either spontaneously or as a result of exposure to chemicals or radiation Such damage to DNA can block replication or transcription, and can result in a high frequency of mutations—consequences that are unacceptable from the standpoint of cell reproduction. To maintain the integrity of their genomes, cells have therefore had to evolve mechanisms to repair damaged DNA. These mechanisms of DNA repair can be divided into two general classes: (1) direct reversal of the chemical reaction responsible for DNA damage, and (2) removal of the damaged bases followed by their replacement with newly synthesized DNA. Where DNA repair fails, additional mechanisms have evolved to enable cells to cope with the damage.

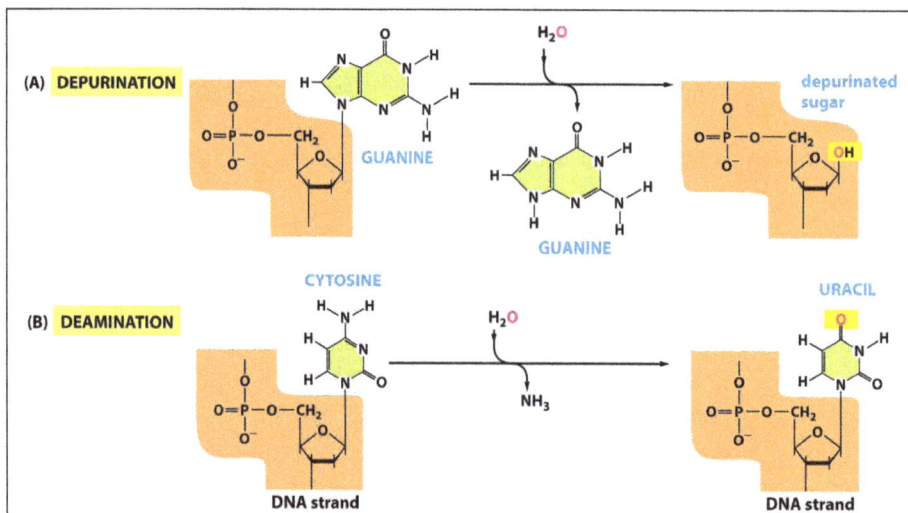

Figure: Spontaneous damage to DNA.

There are two major forms of spontaneous DNA damage: (A) deamination of adenine, cytosine, and guanine, and (B) depurination (loss of purine bases) resulting from cleavage of the bond between the purine bases and deoxyribose, leaving an apurinic (AP) site in DNA. dGMP = deoxyguanosine monophosphate.

(A) UV light induces the formation of pyrimidine dimers, in which two adjacent pyrimidines (e.g., thymines) are joined by a cyclobutane ring structure. (B) Alkylation is the addition of methyl or ethyl groups to various positions on the DNA bases. In this example, alkylation of the $O^6$ position of guanine results in formation of $O^6$-methylguanine. (C) Many carcinogens (e.g., benzo-($a$) pyrene) react with DNA bases, resulting in the addition of large bulky chemical groups to the DNA molecule.

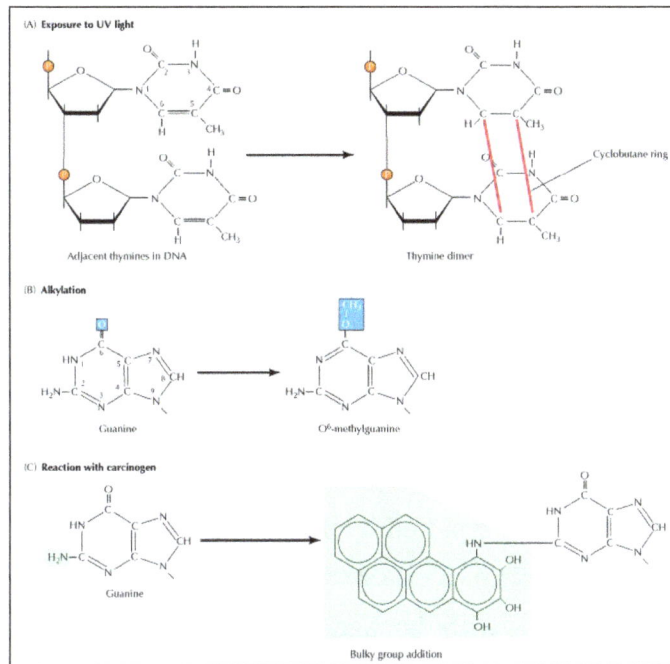

Figure: Examples of DNA damage induced by radiation and chemicals.

## Direct Reversal of DNA Damage

Most damage to DNA is repaired by removal of the damaged bases followed by resynthesis of the excised region. Some lesions in DNA, however, can be repaired by direct reversal of the damage, which may be a more efficient way of dealing with specific types of DNA damage that occur frequently. Only a few types of DNA damage are repaired in this way, particularly pyrimidine dimers resulting from exposure to ultraviolet (UV) light and alkylated guanine residues that have been modified by the addition of methyl or ethyl groups at the $O^6$ position of the purine ring.

UV light is one of the major sources of damage to DNA and is also the most thoroughly studied form of DNA damage in terms of repair mechanisms. Its importance is illustrated by the fact that exposure to solar UV irradiation is the cause of almost all skin cancer in humans. The major type of damage induced by UV light is the formation of pyrimidine dimers, in which adjacent pyrimidines on the same strand of DNA are joined by the formation of a cyclobutane ring resulting from saturation of the double bonds between carbons 5 and 6. The formation of such dimers distorts the structure of the DNA chain and blocks transcription or replication past the site of damage, so their repair is closely correlated with the ability of cells to survive UV irradiation. One mechanism of repairing UV-induced pyrimidine dimers is direct reversal of the dimerization reaction. The process is called photoreactivation because energy derived from visible light is utilized to break the cyclobutane ring structure. The original pyrimidine bases remain in DNA, now restored to their normal state. As might be expected from the fact that solar UV irradiation is a major source of DNA damage for diverse cell types, the repair of pyrimidine dimers by photoreactivation is common to a variety of prokaryotic and eukaryotic cells, including *E. coli*, yeasts, and some species of plants and animals. Curiously, however, photoreactivation is not universal; many species (including humans) lack this mechanism of DNA repair.

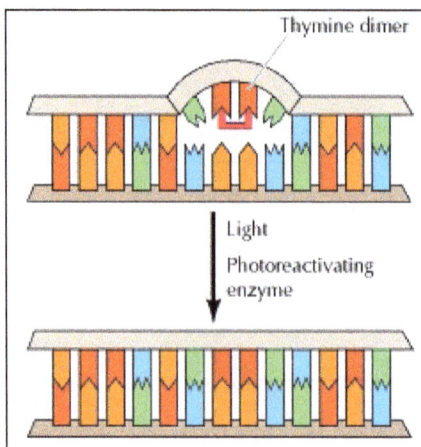

Figure: Direct repair of thymine dimers.

UV-induced thymine dimers can be repaired by photoreactivation, in which energy from visible light is used to split the bonds forming the cyclobutane ring.

Another form of direct repair deals with damage resulting from the reaction between alkylating agents and DNA. Alkylating agents are reactive compounds that can transfer methyl or ethyl groups to a DNA base, thereby chemically modifying the base. A particularly important type of damage is methylation of the $O^6$ position of guanine, because the product, $O^6$-methylguanine, forms complementary base pairs with thymine instead of cytosine. This lesion can be repaired by an enzyme (called $O^6$-methylguanine methyltransferase) that transfers the methyl group from $O^6$-methylguanine to a cysteine residue in its active site. The potentially mutagenic chemical modification is thus removed, and the original guanine is restored. Enzymes that catalyze this direct repair reaction are widespread in both prokaryotes and eukaryotes, including humans.

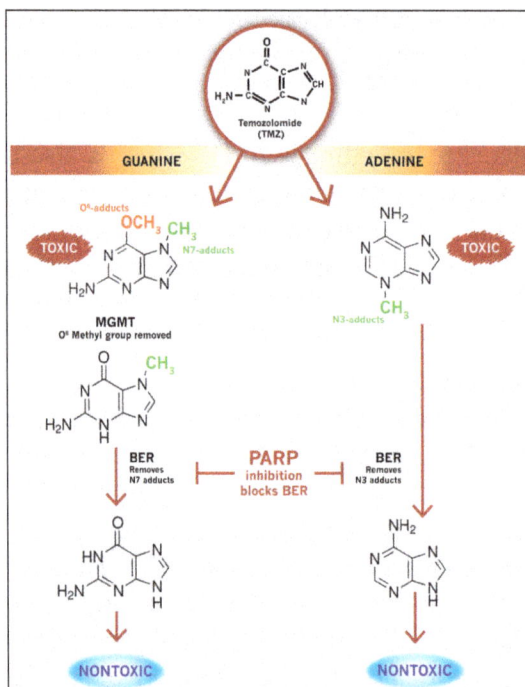

Figure: Repair of $O^6$-methylguanine.

$O^6$-methylguanine methyltransferase transfers the methyl group from $O^6$-methylguanine to a cysteine residue in the enzyme's active site.

## Excision Repair

Although direct repair is an efficient way of dealing with particular types of DNA damage, excision repair is a more general means of repairing a wide variety of chemical alterations to DNA. Consequently, the various types of excision repair are the most important DNA repair mechanisms in both prokaryotic and eukaryotic cells. In excision repair, the damaged DNA is recognized and removed, either as free bases or as nucleotides. The resulting gap is then filled in by synthesis of a new DNA strand, using the undamaged complementary strand as a template. Three types of excision repair—base-excision repair, nucleotide-excision repair, and mismatch repair—enable cells to cope with a variety of different kinds of DNA damage.

The repair of uracil-containing DNA is a good example of base-excision repair, in which single damaged bases are recognized and removed from the DNA molecule. Uracil can arise in DNA by two mechanisms: (1) Uracil (as dUTP [deoxyuridine triphosphate]) is occasionally incorporated in place of thymine during DNA synthesis, and (2) uracil can be formed in DNA by the deamination of cytosine . The second mechanism is of much greater biological significance because it alters the normal pattern of complementary base pairing and thus represents a mutagenic event. The excision of uracil in DNA is catalyzed by DNA glycosylase, an enzyme that cleaves the bond linking the base (uracil) to the deoxyribose of the DNA backbone. This reaction yields free uracil and an apyrimidinic site—a sugar with no base attached. DNA glycosylases also recognize and remove other abnormal bases, including hypoxanthine formed by the deamination of adenine, pyrimidine dimers, alkylated purines other than $O^6$-alkylguanine, and bases damaged by oxidation or ionizing radiation.

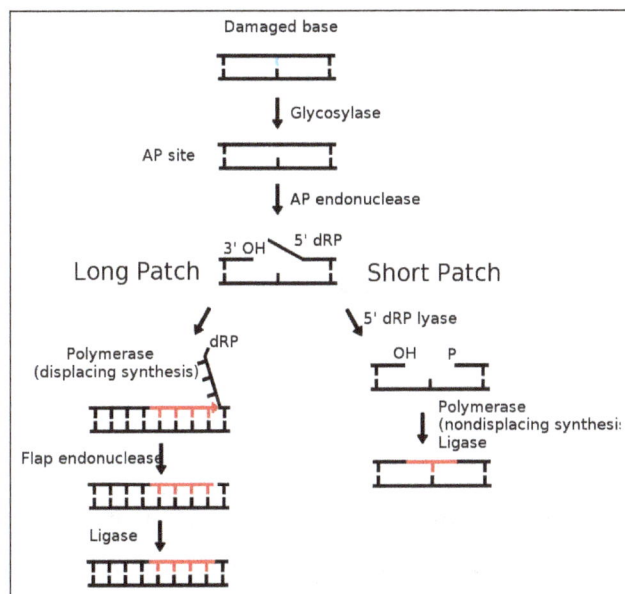

Figure: Base-excision repair.

In this example, uracil (U) has been formed by deamination of cytosine (C) and is therefore opposite a guanine (G) in the complementary strand of DNA. The bond between uracil and the deoxyribose is cleaved by a DNA glycosylase, leaving a sugar with no base attached in the DNA (an

AP site). This site is recognized by AP endonuclease, which cleaves the DNA chain. The remaining deoxyribose is removed by deoxyribosephosphodiesterase. The resulting gap is then filled by DNA polymerase and sealed by ligase, leading to incorporation of the correct base (C) opposite the G.

The result of DNA glycosylase action is the formation of an apyridiminic or apurinic site (generally called an AP site) in DNA. Similar AP sites are formed as the result of the spontaneous loss of purine bases, which occurs at a significant rate under normal cellular conditions. For example, each cell in the human body is estimated to lose several thousand purine bases daily. These sites are repaired by AP endonuclease, which cleaves adjacent to the AP site. The remaining deoxyribose moiety is then removed, and the resulting single-base gap is filled by DNA polymerase and ligase.

Whereas DNA glycosylases recognize only specific forms of damaged bases, other excision repair systems recognize a wide variety of damaged bases that distort the DNA molecule, including UV-induced pyrimidine dimers and bulky groups added to DNA bases as a result of the reaction of many carcinogens with DNA. This widespread form of DNA repair is known as nucleotide-excision repair, because the damaged bases (e.g., a thymine dimer) are removed as part of an oligonucleotide containing the lesion.

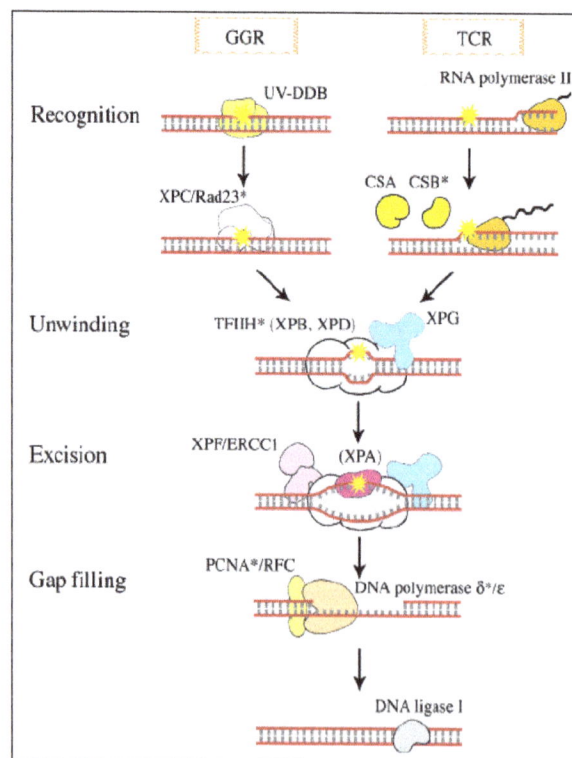

Figure: Nucleotide-excision repair of thymine dimers.

Damaged DNA is recognized and then cleaved on both sides of a thymine dimer by 3' and 5' nucleases. Unwinding by a helicase results in excision of an oligonucleotide containing the damaged bases. The resulting gap is then filled by DNA polymerase and sealed by ligase. In E. coli, nucleotide-excision repair is catalyzed by the products of three genes (uvrA, B, and C) that were identified because mutations at these loci result in extreme sensitivity to UV light. The protein UvrA recognizes damaged DNA and recruits UvrB and UvrC to the site of the lesion. UvrB and UvrC then cleave on the 3' and 5' sides of the damaged site, respectively, thus excising an oligonucleotide consisting of 12 or 13 bases. The UvrABC complex is frequently called an excinuclease, a name that reflects its ability to directly excise an oligonucleotide. The action of a helicase is then required to remove the damage-containing oligonucleotide from the double-stranded DNA molecule, and the resulting gap is filled by DNA polymerase I and sealed by ligase.

Nucleotide-excision repair systems have also been studied extensively in eukaryotes, particularly in yeasts and in humans. In yeasts, as in E. coli, several genes involved in DNA repair (called RAD genes for radiation sensitivity) have been identified by the isolation of mutants with increased sensitivity to UV light. In humans, DNA repair genes have been identified largely by studies of individuals suffering from inherited diseases resulting from deficiencies in the ability to repair DNA damage. The most extensively studied of these diseases is xeroderma pigmentosum (XP), a rare genetic disorder that affects approximately one in 250,000 people. Individuals with this disease are extremely sensitive to UV light and develop multiple skin cancers on the regions of their bodies that are exposed to sunlight. In 1968 James Cleaver made the key discovery that cultured cells from XP patients were deficient in the ability to carry out nucleotide-excision repair. This observation not only provided the first link between DNA repair and cancer, but also suggested the use of XP cells as an experimental system to identify human DNA repair genes. The identification of human DNA repair genes has been accomplished by studies not only of XP cells, but also of two other human diseases resulting from DNA repair defects (Cockayne's syndrome and trichothiodystrophy) and of UV-sensitive mutants of rodent cell lines. The availability of mammalian cells with defects in DNA repair has allowed the cloning of repair genes based on the ability of wild-type alleles to restore normal UV sensitivity to mutant cells in gene transfer assays, thereby opening the door to experimental analysis of nucleotide-excision repair in mammalian cells.

Molecular cloning has now identified seven different repair genes (designated XPA through XPG) that are mutated in cases of xeroderma pigmentosum, as well as in some cases of Cockayne's syndrome, trichothiodystrophy, and UV-sensitive mutants of rodent cells. Table lists the enzymes encoded by these genes. Some UV-sensitive rodent cells have mutations in yet another repair gene, called ERCC1 (for excision repair cross complementing), which has not been found to be mutated in known human diseases. It is notable that the proteins encoded by these human DNA repair genes are closely related to proteins encoded by yeast RAD genes, indicating that nucleotide-excision repair is highly conserved throughout eukaryotes.

Table: Enzymes Involved in Nucleotide-Excision Repair.

| Human | Yeast | Function |
|---|---|---|
| XPA | RAD14 | Damage recognition |
| XPB | RAD25 | Helicase |
| XPC | RAD4 | DNA binding |
| XPD | RAD3 | Helicase |
| XPF | RAD1 | 5' nuclease |
| XPG | RAD2 | 3' nuclease |
| ERCC1 | RAD10 | Dimer with XPF |

With cloned yeast and human repair genes available, it has been possible to purify their encoded proteins and develop in vitro systems to study the repair process. Although some steps remain to be fully elucidated, these studies have led to the development of a basic model for nucleotide-excision repair in eukaryotic cells. In mammalian cells, the XPA protein (and possibly also XPC) initiates repair by recognizing damaged DNA and forming complexes with other proteins involved in the repair process. These include the XPB and XPD proteins, which act as helicases that unwind the damaged DNA. In addition, the binding of XPA to damaged DNA leads to the recruitment of XPF (as a heterodimer with ERCC1) and XPG to the repair complex. XPF/ERCC1 and XPG are endonucleases, which cleave DNA on the 5′ and 3′ sides of the damaged site, respectively. This cleavage excises an oligonucleotide consisting of approximately 30 bases. The resulting gap then appears to be filled in by DNA polymerase δ or ε (in association with RFC and PCNA) and sealed by ligase.

An intriguing feature of nucleotide-excision repair is its relationship to transcription. A connection between transcription and repair was first suggested by experiments showing that transcribed strands of DNA are repaired more rapidly than non-transcribed strands in both E. coli and mammalian cells. Since DNA damage blocks transcription, this transcription-repair coupling is thought to be advantageous by allowing the cell to preferentially repair damage to actively expressed genes. In E. coli, the mechanism of transcription-repair coupling involves recognition of RNA polymerase stalled at a lesion in the DNA strand being transcribed. The stalled RNA polymerase is recognized by a protein called transcription-repair coupling factor, which displaces RNA polymerase and recruits the UvrABC excinuclease to the site of damage.

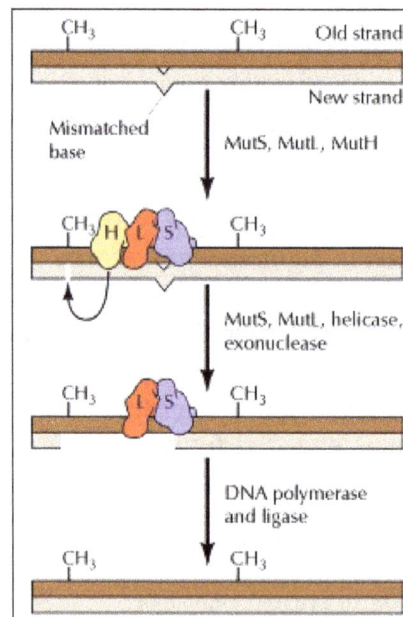

Figure: Mismatch repair in E. coli.

Although the molecular mechanism of transcription-repair coupling in mammalian cells is not yet known, it is noteworthy that the XPB and XPD helicases are components of a multisubunit transcription factor (called TFIIH) that is required to initiate the transcription of eukaryotic genes. Thus, these helicases appear to be required for the unwinding of DNA during both transcription and nucleotide-excision repair, providing a direct biochemical link between these two processes. Patients suffering from Cockayne's syndrome are also characterized from a failure to preferentially repair transcribed DNA strands, suggesting that the proteins encoded by the two genes known to be responsible for this disease (CSA and CSB) function in transcription-coupled repair.

In addition, one of the genes responsible for inherited breast cancer in humans (BRCA1) appears to encode a protein specifically involved in transcription-coupled repair of oxidative DNA damage, suggesting that defects in this type of DNA repair can lead to the development of one of the most common cancers in women.

A third excision repair system recognizes mismatched bases that are incorporated during DNA replication. Many such mismatched bases are removed by the proofreading activity of DNA polymerase. The ones that are missed are subject to later correction by the mismatch repair system, which scans newly replicated DNA. If a mismatch is found, the enzymes of this repair system are able to identify and excise the mismatched base specifically from the newly replicated DNA strand, allowing the error to be corrected and the original sequence restored.

In E. coli, the ability of the mismatch repair system to distinguish between parental DNA and newly synthesized DNA is based on the fact that DNA of this bacterium is modified by the methylation of adenine residues within the sequence GATC to form 6-methyladenine. Since methylation occurs after replication, newly synthesized DNA strands are not methylated and thus can be specifically recognized by the mismatch repair enzymes. Mismatch repair is initiated by the protein MutS, which recognizes the mismatch and forms a complex with two other proteins called MutL and MutH. The MutH endonuclease then cleaves the unmethylated DNA strand at a GATC sequence. MutL and MutS then act together with an exonuclease and a helicase to excise the DNA between the strand break and the mismatch, with the resulting gap being filled by DNA polymerase and ligase.

The mismatch repair system detects and excises mismatched bases in newly replicated DNA, which is distinguished from the parental strand because it has not yet been methylated. MutS binds to the mismatched base, followed by MutL. The binding of MutL activates MutH, which cleaves the unmodified strand opposite a site of methylation. MutS and MutL, together with a helicase and an exonuclease, then excise the portion of the unmodified strand that contains the mismatch. The gap is then filled by DNA polymerase and sealed by ligase.

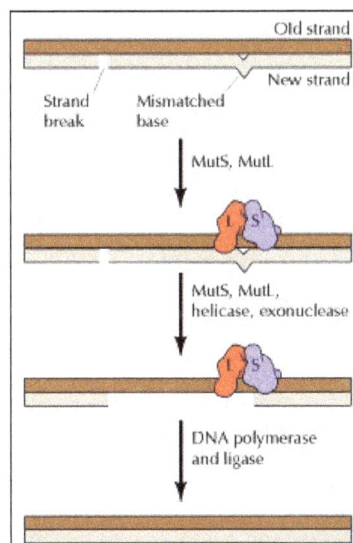

Figure: Mismatch repair in mammalian cells.

Eukaryotes have a similar mismatch repair system, although the mechanism by which eukaryotic cells identify newly replicated DNA differs from that used by E. coli. In mammalian cells, it appears that the strand-specificity of mismatch repair is determined by the presence of single-strand breaks (which would be present in newly replicated DNA) in the strand to be repaired. The eukaryotic homologs of MutS and MutL then bind to the mismatched base and direct excision of the DNA between the strand break and the mismatch, as in E. coli. The importance of this repair system is dramatically illustrated by the fact that mutations in the human homologs of MutS and MutL are responsible for a common type of inherited colon cancer (hereditary nonpolyposis colorectal cancer, or HNPCC). HNPCC is one of the most common inherited diseases; it affects as many as one in 200 people and is responsible for about 15% of all colorectal cancers in this country. The relationship between HNPCC and defects in mismatch repair was discovered in 1993, when two groups of researchers cloned the human homolog of MutS and found that mutations in this gene were responsible for about half of all HNPCC cases. Subsequent studies have shown that most of the remaining cases of HNPCC are caused by mutations in one of three human genes that are homologs of MutL.

Mismatch repair in mammalian cells is similar to E. coli, except that the newly replicated strand is distinguished from the parental strand because it contains strand breaks. MutS and MutL bind to the mismatched base and direct excision of the DNA between the strand break and the mismatch.

## Postreplication Repair

The direct reversal and excision repair systems act to correct DNA damage before replication, so that replicative DNA synthesis can proceed using an undamaged DNA strand as a template. Should these systems fail, however, the cell has alternative mechanisms for dealing with damaged DNA at the replication fork. Pyrimidine dimers and many other types of lesions cannot be copied by the normal action of DNA polymerases, so replication is blocked at the sites of such damage. Downstream of the damaged site, however, replication can be initiated again by the synthesis of an Okazaki fragment and can proceed along the damaged template strand. The result is a daughter strand that has a gap opposite the site of damage to the parental strand. One of two types of mechanisms may be used to repair such gaps in newly synthesized DNA: recombination repair or error-prone repair.

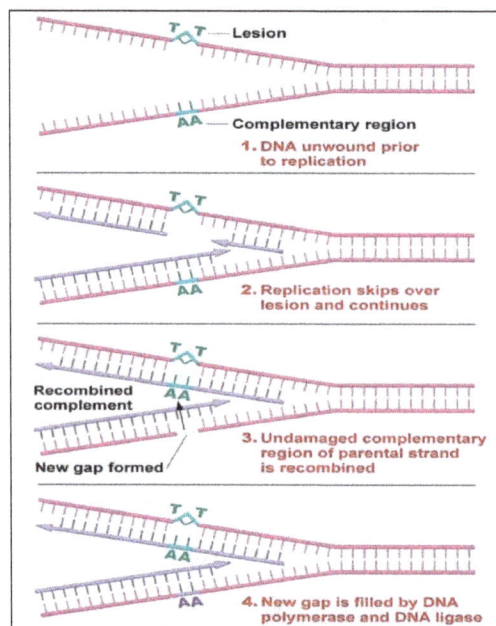

Figure: Postreplication repair.

The presence of a thymine dimer blocks replication, but DNA polymerase can bypass the lesion and reinitiate replication at a new site downstream of the dimer. The result is a gap opposite the dimer in the newly synthesized DNA strand. In recombinational repair, this gap is filled by recombination with the undamaged parental strand. Although this leaves a gap in the previously intact parental strand, the gap can be filled by the actions of polymerase and ligase, using the intact daughter strand as a template. Two intact DNA molecules are thus formed, and the remaining thymine dimer eventually can be removed by excision repair.

Recombinational repair depends on the fact that one strand of the parental DNA was undamaged and therefore was copied during replication to yield a normal daughter molecule. The undamaged parental strand can be used to fill the gap opposite the site of damage in the other daughter molecule by recombination between homologous DNA sequences. Because the resulting gap in the previously intact parental strand is opposite an undamaged strand, it can be filled in by DNA polymerase. Although the other parent molecule still retains the original damage (e.g., a pyrimidine dimer), the damage now lies opposite a normal strand and can be dealt with later by excision repair. By a similar mechanism, recombination with an intact DNA molecule can be used to repair double strand breaks, which are frequently introduced into DNA by radiation and other damaging agents.

In error-prone repair, a gap opposite a site of DNA damage is filled by newly synthesized DNA. Since the new DNA is synthesized from a damaged template strand, this form of DNA synthesis is very inaccurate and leads to frequent mutations. It is used only in bacteria that have been subjected to potentially lethal conditions, such as extensive UV irradiation. Such treatments induce the SOS response, which may be viewed as a mechanism for dealing with extreme environmental stress. The SOS response includes inhibition of cell division and induction of repair systems to cope with a high level of DNA damage. Under these conditions, error-prone repair mechanisms are used, presumably as a way of dealing with damage so extensive that cell death is the only alternative.

## Genetic Recombination

DNA recombination involves the exchange of genetic material either between multiple chromosomes or between different regions of the same chromosome. This process is generally mediated by homology; that is, homologous regions of chromosomes line up in preparation for exchange, and some degree of sequence identity is required. Various cases of nonhomologous recombination do exist, however.

One important instance of recombination in diploid eukaryotic organisms is the exchange of genetic information between newly duplicated chromosomes during the process of meiosis. In this instance, the outcome of recombination is to ensure that each gamete includes both maternally and paternally derived genetic information, such that the resulting offspring will inherit genes from all four of its grandparents, thereby acquiring a maximum amount of genetic diversity. Recombination is also used in DNA repair (particularly in the repair of double-stranded breaks), as well as during DNA replication to assist in filling gaps and preventing stalling of the replication fork. In these cases, a sister chromatid serves as the donor of missing material via recombination followed by DNA synthesis.

The role of recombination during the inheritance of chromosomes was first demonstrated through experiments with maize. Specifically, in 1931, Barbara McClintock and Harriet Creighton obtained evidence for recombination by physically tracking an unusual knob structure within certain maize chromosomes through multiple genetic crosses. Using a strain of maize in which one member of a chromosome pair exhibited the knob but its homologue did not, the scientists were able to show that some alleles were physically linked to the knobbed chromosome, while other alleles were tied to the normal chromosome. McClintock and Creighton then followed these alleles through meiosis, showing that alleles for specific phenotypic traits were physically exchanged between chromosomes. Evidence for this finding came from the fact that alleles first introduced into the cross on a knobbed chromosome later appeared in offspring without the knob; similarly, alleles initially introduced on a knobless chromosome subsequently appeared in progeny with the knob.

Recombination also occurs in prokaryotic cells, and it has been especially well characterized in *E. coli*. Although bacteria do not undergo meiosis, they do engage in a type of sexual reproduction called conjugation, during which genetic material is transferred from one bacterium to another and may be recombined in the recipient cell. As in eukaryotes, recombination also plays important roles in DNA repair and replication in prokaryotic organisms.

## Models of Recombination

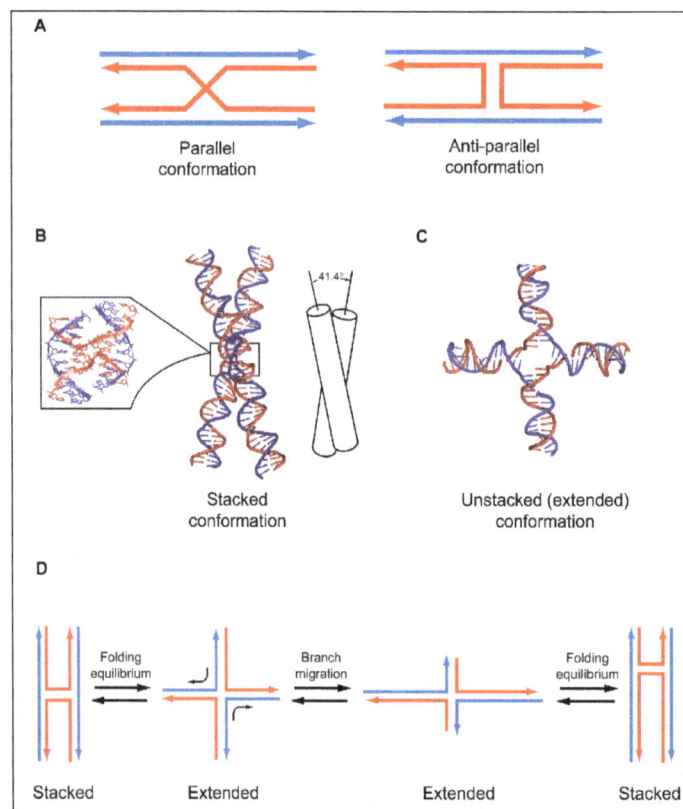

Figure: Structure of the Holliday junction.

Although common, genetic recombination is a highly complex process. It involves the alignment of two homologous DNA strands (the requirement for homology suggests that this occurs through complementary base-pairing, but this has not been definitively shown), precise breakage of each strand, exchange between the strands, and sealing of the resulting recombined molecules. This process occurs with a high degree of accuracy at high frequency in both eukaryotic and prokaryotic cells.

The basic steps of recombination can occur in two pathways, according to whether the initial break is single or double stranded. In the single-stranded model, following the alignment of homologous chromosomes, a break is introduced into one DNA strand on each chromosome, leaving two free ends. Each end then crosses over and invades the other chromosome, forming a structure called a Holliday junction. The next step, called branch migration, takes place as the junction travels down the DNA. The junction is then resolved either horizontally, which produces no recombination, or vertically, which results in an exchange of DNA. In the alternate pathway initiated by double-stranded breaks, the ends at the breakpoints are converted into single strands by the addition of 3' tails. These ends can then perform strand invasion, producing two Holliday junctions. From that point forward, resolution proceeds as in the single-stranded model (Note that a third model of recombination, synthesis-dependent strand annealing [SDSA], has also been proposed to account for the lack of crossover typical of recombination in mitotic cells and observed in some meiotic cells to a lesser degree.)

## Recombination Enzymes

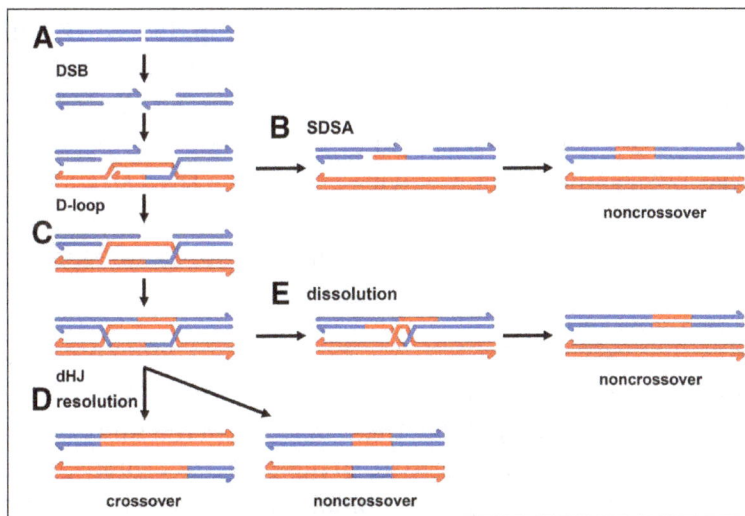

Figure: Repair of DNA double-strand breaks by DSBR and SDSA.

Double-strand breaks (DSBs) can be repaired by several homologous recombination (HR)-mediated pathways, including double-strand break repair (DSBR) and synthesis-dependent strand annealing (SDSA). a) In both pathways, repair is initiated by resection of a DSB to provide 3' single-stranded DNA (ssDNA) overhangs. Strand invasion by these 3' ssDNA overhangs into a homologous sequence is followed by DNA synthesis at the invading end. b) After strand invasion and synthesis, the second DSB end can be captured to form an intermediate with two Holliday junctions (HJs). After gap-repair DNA synthesis and ligation, the structure is resolved at the HJs in a non-crossover (black arrow heads at both HJs) or crossover mode (green arrow heads at one HJ and black arrow heads at the other HJ). c) Alternatively, the reaction can proceed to SDSA by strand displacement, annealing of the extended single-strand end to the ssDNA on the other break end, followed by gap-filling DNA synthesis and ligation. The repair product from SDSA is always non-crossover.

No matter which pathway is used, a number of enzymes are required to complete the steps of recombination. The genes that code for these enzymes were first identified in E. coli by the isolation of mutant cells that were deficient in recombination. This research revealed that the recA gene encodes a protein necessary for strand invasion. Meanwhile, the recB, recC, and recD genes code for three polypeptides that join together to form a protein complex known as RecBCD; this complex has the capacity to unwind double-stranded DNA and cleave strands. Two other genes, ruvA and ruvB, encode enzymes that catalyze branch migration, while Holliday structures are resolved by the protein resolvase, which is product of the ruvC gene. Several enzymes involved in DNA replication, such as ligase and DNA polymerase, also contribute to recombination.

In eukaryotes, recombination has been perhaps most thoroughly studied in the budding yeast Saccharomyces cerevisiae. Many of the enzymes identified in this yeast have also been found in other organisms, including mammalian cells. Such studies reveal that the Rad genes (named for the fact that their activity was found to be sensitive to radiation) play a key role in eukaryotic recombination. In particular, the Rad51 gene, which is homologous to recA, encodes a protein (called Rad51) that has recombinase activity. This gene is highly conserved, but the accessory proteins that assist Rad51 appear to vary among organisms. For example, the Rad52 protein is found in both yeast and humans, but it is missing in Drosophila melanogaster and C. elegans.

In eukaryotic cells, single-stranded DNA (ssDNA) becomes rapidly coated with the protein RPA (replication protein A). RPA has a higher affinity for ssDNA than Rad51, and it therefore can inhibit recombination by blocking Rad51's access to the single strand needed for invasion. In yeast, however, binding of Rad51 to ssDNA is enhanced by the proteins Rad52 and the complex Rad55-Rad57. Once access has been gained, Rad51 polymerizes on the DNA strand to form what is called a presynaptic filament, which is a right-handed helical filament containing six Rad51 molecules and 18 nucleotides per helical repeat. The search for DNA homology and formation of the junction between homologous regions is then carried out within the catalytic center of the filament.

In addition to proteins that assist Rad51 activity, there are also some proteins that inhibit it. In yeast, for instance, the helicase Srs2 dismantles the Rad51-ssDNA complex, while the proteins Sgs1 and BLM inhibit the complex. It is thought that these proteins play a role in preventing recombination during DNA replication when it is not needed. In humans, the tumor suppressor genes BRCA1 and BRCA2 also play a role in regulating recombination. Individuals who are heterozygous for BRCA2 are subject to increased risk for breast and ovarian cancer; loss of both alleles causes Fanconi's anemia, a genetic disease characterized by predisposition to cancer, among other defects. BRCA2 appears to promote Rad51 binding to ssDNA.

## Homologous Sequences brought together

The enzymes and mechanisms that carry out the process of homologous recombination are fairly well delineated. Not so well understood is the important question of how homologous sequences come to be in proximity so that recombination can proceed. In their 2008 review, barzel and kupiec describe two alternate hypotheses, one of which they call the null model. This model proposes that homologues find one another through a passive process of diffusion, in which the DNA sequence at the broken end of a strand is sequentially compared to all of the other potential end sequences in the genome. In order for diffusion to account for the rapid repair of double-stranded breaks observed in yeast, however, Barzel and Kupiec calculate that each homology search would have to proceed at a speed 40 times faster than the rate at which DNA polymerase adds a single nucleotide to a replicating DNA chain, which seems unlikely.

An alternate hypothesis proposes that homologous chromosomes reside in pairs constitutively. Acting against this hypothesis is the finding that in induced recombination experiments, the broken ends of strands recombine with what are called ectopic homologues (areas of fortuitous sequence identity) as frequently as they recombine with their true homologous chromosomes. Furthermore, although homologous pairing has been observed in somatic cells of some organisms (e.g., Drosophila, Neurospora), it is not widely seen in the cells of other organisms, including mammals. As Barzel and Kupiec point out, the absence of general homologous pairing does not necessarily mean random assortment. Instead, discrete sections of chromosomes may be required for homology. The use of subdomains for homology searches would reduce the time it takes to find a homologous partner. Despite such theories, the exact mechanism responsible for locating and lining up homologous segments remains to be determined.

## RNA Interference

RNA silencing is a novel gene regulatory mechanism that limits the transcript level by either suppressing transcription (transcriptional gene silencing [TGS]) or by activating a sequence-specific RNA degradation process (posttranscriptional gene silencing [PTGS]/RNA interference [RNAi]). Although there is a mechanistic connection between TGS and PTGS, TGS is an emerging field while PTGS is undergoing an explosion in its information content. Here, we have limited our discussion to PTGS/RNAi-related phenomena.

Pioneering observations on PTGS/RNAi were reported in plants, but later on RNAi-related events were described in almost all eukaryotic organisms, including protozoa, flies, nematodes, insects, parasites, and mouse and human cell lines. Three phenotypically different but mechanistically similar forms of RNAi, cosuppression or PTGS in plants, quelling in fungi, and RNAi in the animal kingdom, have been described. More recently, micro-RNA formation, heterochromatinization, etc., have been revealed as other facets of naturally occurring RNAi processes of eukaryotic cells.

During the occurrence of RNAi/PTGS, double-stranded RNA (dsRNA) molecules, which cleave the inducer molecules into smaller pieces first and eventually destroy the cellular or viral cognate mRNA molecules (called the target) act as inducers or activators of this process. As a result, the target mRNAs cannot accumulate in the cytosol, although they remain detectable by nuclear run-on assays. In certain instances, the DNA expressing the target mRNA also undergoes methylation as a by-product of the degradation process. The natural functions of RNAi and its related processes seem to be protection of the genome against invasion by mobile genetic elements such as viruses and transposons as well as orchestrated functioning of the developmental programs of eukaryotic organisms. There are several excellent recent reviews which deal with different aspects of RNAi separately. Here, we have put together the various aspects of the RNAi process known to date, identified the mechanistic similarities and differences operating in various forms of eukaryotic life, and focused on the experimental results that have led to conceptual advancements in this field.

## Unraveling RNA Silencing

In order to understand the process of homology-dependent RNA silencing, it would be prudent to overview the process itself and describe its important features. In the later part of this review, the genetics, biochemistry, and potential therapeutic applications of the process will be dealt with.

## PTGS in Plants

In plants, the RNA silencing story unfolded serendipitously during a search for transgenic petunia flowers that were expected to be purpler. In 1990, R. Jorgensen's laboratory wanted to upregulate the activity of a gene for chalcone synthase (chsA), an enzyme involved in the production of anthocyanin pigments. Surprisingly, some of the transgenic petunia plants harboring the *chsA* coding region under the control of a 35S promoter lost both endogene and transgene chalcone synthase activity, and thus many of the flowers were variegated or developed white sectors. The loss of cytosolic *chsA* mRNA was not associated with reduced transcription, as demonstrated by run-on transcription tests in isolated nuclei. Jorgensen coined the term cosuppression to describe the loss of mRNAs of both the endo- and the transgene.

Around the same time, two other laboratories also reported that introduction of the transcribing-sense transgenes could downregulate the expression of homologous endogenous genes. Subsequently, many similar events of cosuppression were reported in the literature. All cases of cosuppression resulted in the degradation of endogene and transgene RNAs after nuclear transcription had occurred. Since posttranscriptional RNA degradation was observed in a wide range of transgenes expressing the plant, bacterial, or viral sequences, it was rechristened posttranscriptional gene silencing (PTGS). PTGS could be initiated not only by sense transgenes but also by antisense transgenes, and biochemical evidence suggests that similar mechanisms might operate in both cases. It is worthwhile to point out that although the cosuppression phenomenon was originally observed in plants, it is not restricted to plants and has also been demonstrated in metazoans and mammals.

In keeping with the times, the observed alterations in the PTGS-related phenotypes were attributed to multiple-site integrations, aberrant RNA formations, repeat structures of the transgenes, etc. Later on, it became clear that the expression of the transgene led to the formation of dsRNA, which, in turn, initiated PTGS. For example, in the case of cosuppressed petunia plants, chsA mRNA formed a partial duplex, since there are regions of self-complementarity located between chsA 3' coding region and its 3' untranslated region. This was revealed by DNA sequence analysis and experimental detection of in vitro-transcribed, RNase-resistant duplex chsA RNA. In an independent study, a p35S-ACC (1-aminocyclopropane-1-carboxylate [ACC] oxidase) sense transgene carrying a small inverted repeat in the 5' untranslated region was introduced into tomato to test the role of dsRNA structure as an inducer of PTGS. Cosuppression of the endogenous acc gene occurred at a higher frequency in these plants than in those harboring only the p35S-ACC sense transgene without the inverted repeat.

Reports from several laboratories in the past few years have established that the loss in steady-state accumulation of the target mRNA is almost total if the designed transgene construct of the transgenic plant produces the nuclear transcript in the duplex conformation. Very recently it was reported that the expression of self-cRNA of plum pox virus under the control of rolC promoter caused degradation of transgenic viral RNA and as a result, the systemic disease resistance to challenge inoculum of plum pox virus occurred with a high frequency in transgenic Nicotiana benthamiana. This evidence points out that the production of dsRNA is required to initiate PTGS in plants. Based on this, plants carrying strongly transcribing transgenes in both the sense and antisense orientations are currently being produced that show strong PTGS features. These transgenic plants can silence endogene, invading viral RNA, or unwanted foreign genes in a sequence-specific and heritable manner.

Generally, the sense and antisense components of the above-mentioned transgenes are separated only by an intron to increase the efficacy of PTGS. For example, Arabidopsis thaliana and Lycopersicon esculentum (tomato) plants were transformed with a transgene construct designed to generate self-complementary iaaM and ipt transcripts. iaaM and ipt are oncogenes of agrobacteria that are responsible for crown gall formation in infected plants. The transgenic lines retained susceptibility to Agrobacterium transformation but were highly refractory to tumorigenesis, providing functional resistance to crown gall disease by posttranscriptional degradation of the iaaM and ipt transcripts.

## Quelling and RNAi

While reports of PTGS in plants were piling up, homology-dependent gene silencing phenomena were also observed independently in fungal systems. These events were called quelling. Quelling came to light during attempts to boost the production of an orange pigment made by the gene al1 of the fungus Neurospora crassa. An N. crassa strain containing a wild-type al1+ gene (orange phenotype) was transformed with a plasmid containing a 1,500-bp fragment of the coding sequence of the al1 gene. A few transformants were stably quelled and showed albino phenotypes. In the al1-quelled strain, the level of unspliced al1 mRNA was similar to that of the wild-type strain, whereas the native al1 mRNA was highly reduced, indicating that quelling and not the rate of transcription affected the level of mature mRNA in a homology-dependent manner.

The phenomenon of RNAi first came into the limelight following the discovery by Fire et al., who unequivocally demonstrated the biochemical nature of inducers in gene silencing by introducing purified dsRNA directly into the body of Caenorhabditis elegans. The investigators injected dsRNA corresponding to a 742-nucleotide segment of unc22 into either the gonad or body cavity region of an adult nematode. unc22 encodes an abundant but nonessential myofilament protein, and the decrease in unc22 activity is supposed to produce an increasingly severe twitching phenotype. The injected animal showed weak twitching, whereas the progeny individuals were strong twitchers. The investigators showed that similar loss-of-function individuals could also be generated with dsRNAs corresponding to four other nematode genes. The phenotypes produced by interference by various dsRNAs were extremely specific.

This experiment paved the way for easy production of null mutants, and the process of silencing a functional gene by exogenous application of dsRNA was termed RNA interference (RNAi). RNAi in *C. elegans* was also initiated simply by soaking the worms in a solution containing dsRNAs or by feeding the worms *Escherichia coli* organisms that expressed the dsRNAs. This is a very potent method, requiring only catalytic amounts of dsRNA per cell to silence gene expression. The silencing spread not only from the gut of the worm to the remainder of the body, but also through the germ line to several generations. These phenomena of RNAi have also been demonstrated to occur in *Drosophila melanogaster* and many other invertebrates and vertebrates.

## Virus-infected Plants: Virus-induced Gene Silencing

Besides the processes mentioned above, homology-driven RNA degradation also occurs during the growth of viral genomes in infected plants. Viruses can be either the source, the target, or both the source and the target of silencing. PTGS mediated by viruses can occur with RNA viruses, which replicate in the cytoplasm, and also with DNA viruses, which replicate in the nucleus. As early as in the 1920s, it was known that plants could be protected from a severe virus by prior infection with a mild strain of a closely related virus. Although the mechanism of such cross protection in plants remained unknown for a long time, such phenomena could be explained partly in terms of PTGS that could be induced by the mild strain and targeted later against the virulent viral genome. It was also found that transforming plants with virus-derived transgenes gave protection against the challenge viruses even when no transgene protein was produced.

Analyses of these virus-resistant plants revealed that the transgenes were highly transcribed in the nucleus, whereas the steady-state level of cytoplasmic mRNA was very low. Further analysis suggested that some of the transgenic mRNA molecules assumed the conformation of dsRNA, which triggered sequence-specific degradation of self and other homologous or cRNA sequences in the cytoplasm. Thus, in the virus-resistant lines, not only the transgene mRNAs but also the mRNA from the homologous endogenous gene and the invading viral RNA (with homology to the transgene) were degraded.

Another form of virus-induced gene silencing is the phenomenon of viral recovery itself. When *Brassica napus* was inoculated with cauliflower mosaic virus (a DNA virus), lesions at the site of virus entry were visible 5 to 7 days postinoculation. Symptoms of systemic infections were apparent by 10 to 14 days postinoculation. Symptoms were most prominent at 30 to 40 days postinoculation and declined thereafter (i.e., the plants recovered), with the newly emergent leaves remaining asymptomatic at 50 days postinoculation.

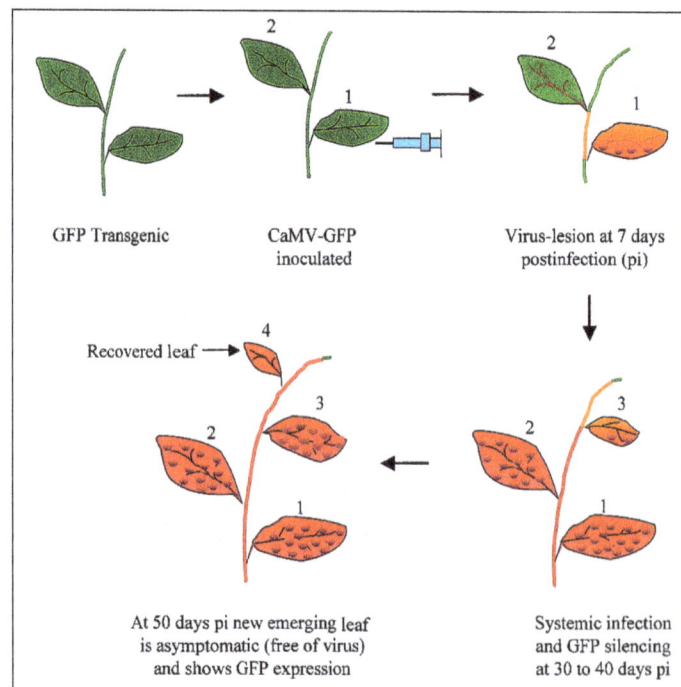

GFP Transgenic

CaMV-GFP
inoculated

Virus-lesion at 7 days
postinfection (pi)

Recovered leaf

At 50 days pi new emerging leaf
is asymptomatic (free of virus)
and shows GFP expression

Systemic infection
and GFP silencing
at 30 to 40 days pi

Diagrammatically illustrates the systemic spread of RNAi in plants. Such recovery occurred by a PTGS-like mechanism because 19S and 35S RNAs encoded by the cauliflower mosaic virus were degraded while cauliflower mosaic virus DNA was still replicating in the nucleus. Induction of PTGS was visualized if the cauliflower mosaic virus infection and subsequent recovery were followed up in a transgenic *B. napus* expressing a p35S-GUS (β-glucuronidase) transgene. At the site of inoculation, GUS silencing associated with local lesions was first observed 7 days postinoculation. GUS silencing eventually spread systemically, and the GUS activity of the entire plant was suppressed by 50 days postinoculation. In this particular example, cauliflower mosaic virus acted as the inducer of PTGS for the transgenes sharing homology with the virus within the transcribed region. However, the virus itself was also the target of the induced PTGS, since 19S and 35S RNAs were found degraded.

Schematic illustration of systemic viral spread as well as RNAi and subsequent viral recovery in plants. Green and red indicate the presence and loss of GFP fluorescence, respectively, and orange denotes the presence of both colors. The red dots on leaves show viral lesions. The bold arrows indicate the stages of plant growth, and the leaves are numbered accordingly. An arrow with a thin line shows a newly emerged leaf recovered from viral attack.

A similar example of virus-induced gene silencing was found when *Nicotiana clevelandii* was infected with an RNA nepovirus, tomato black ring virus. RNA viruses make abundant dsRNA during intracellular replication of their genomes and thus elicit cellular PTGS degradative activity. Virus-induced gene silencing also occurs with viruses that do not undergo recovery. When a DNA geminivirus, tomato golden mosaic virus (TGMV), infected *N. benthamiana*, a high level of viral DNA replication in the nucleus and accumulation of viral RNA in the cytoplasm occurred. An infection by a recombinant TGMV carrying the coding sequence of the sulfur (*su*) gene of the host plant in either the sense or antisense orientation led to the bleaching of leaves due to PTGS of the endogenous *su* gene, but the DNA of the recombinant did not fail to replicate. Here, TGMV acted as an inducer of PTGS but was not itself a target of PTGS. Thus, plant viruses elicit PTGS but sometimes can escape the degradative PTGS activity.

Based on the principles of virus-induced gene silencing, vectors designed with the genome sequence of RNA virus's tobacco mosaic virus, potato virus X, and tobacco rattle virus are being widely used to knock down the expression of host genes. The characteristics of many plant genes were revealed by observing the loss-of-function-related phenotypic changes when the recombinant vectors incorporating the concerned host genes were introduced into plants. Of these vectors, the TRV-based are more promising because these are capable of inducing-meristematic gene silencing, which has not been possible to achieve with other RNA virus-based vectors. Meristematic gene silencing employing TGMV vectors has also been reported. Thus, virus-induced gene silencing-based techniques are extremely useful for studies related to functional genomics in plants.

## Important Features of RNA Silencing

Independently of one another, investigations on diverse organisms, labeled variously as PTGS in plants, RNAi in animals, quelling in fungi, and virus-induced gene silencing, have converged on a universal paradigm of gene regulation. The critical common components of the paradigm are that (i) the inducer is the dsRNA, (ii) the target RNA is degraded in a homology-dependent fashion, and, as we will see later, (iii) the degradative machinery requires a set of proteins which are similar in structure and function across most organisms. In most of these processes, certain invariant features are observed, including the formation of small interfering RNA (siRNA) and the organism-specific systemic transmission of silencing from its site of initiation.

## siRNA

The key insight in the process of PTGS was provided from the experiments of Baulcombe and Hamilton, who identified the product of RNA degradation as a small RNA species (siRNA) of ≈25 nucleotides of both sense and antisense polarity. siRNAs are formed and accumulate as double-stranded RNA molecules of defined chemical structures, as mentioned later. siRNAs were detected first in plants undergoing either suppression or virus-induced gene silencing and were not detectable in control plants that were not silenced. siRNAs were subsequently discovered in *Drosophila* tissue culture cells in which RNAi was induced by introducing >500-nucleotide-long exogenous dsRNA, in *Drosophila* embryo extracts that were carrying out RNAi in vitro, and also in *Drosophila* embryos that were injected with dsRNA. Thus, the generation of siRNA (21 to 25 nucleotides) turned out to be the signature of any homology-dependent RNA-silencing event.

The siRNAs resemble breakdown products of an *E. coli* RNase III-like digestion. In particular, each strand of siRNA has 5′-phosphate and 3′-hydroxyl termini and 2- to 3-nucleotide 3′ overhangs. Interestingly, in vitro-synthesized siRNAs can, in turn, induce specific RNA degradation when added exogenously to *Drosophila* cell extracts. Specific inhibition of gene expression by these siRNAs has also been observed in many invertebrate and some vertebrate systems. Recently, Schwarz et al. provided direct biochemical evidence that the siRNAs could act as guide RNAs for cognate mRNA degradation.

## Amplification and Systemic Transmission

Besides the formation of siRNAs, another intriguing characteristic of homology-dependent gene silencing is that the inducer dsRNA molecules do not act stoichiometrically. It was estimated that only two molecules of dsRNA per cell were able to induce RNAi of an abundantly expressed *C. elegans* gene such as *unc22*. In another report, injection of dsRNA into the intestine of a *C. elegans* hermaphrodite generated RNAi, which could be stably inherited to the $F_2$ generation. These two findings led to the proposal that RNAi signals could be systemic and amplifiable in nature. The similar systemic effects of RNAi have also been demonstrated in the planarian *Schmidtea mediterranea* and the cnidarian *Hydra magnipapillata*.

Similar evidence is also available for plant PTGS. The new tissues growing from a GUS-expressing scion grafted onto a GUS-silenced rootstock show progressive silencing of GUS expression. The silencing signal seems to spread by a nonmetabolic, gene-specific diffusible signal, which travels both between cells, through plasmadesmata, and long distances via the phloem. In the case of virus-induced gene silencing, the systemic character has also been revealed. To account for the gene specificity of a systemic signal, it has been proposed that the signal could be an RNA molecule. However, such processes are not universal, as these are not found in flies and mammals.

## Components of Gene Silencing

Both genetic and biochemical approaches have been undertaken to understand the basis of silencing. Genetic screens were carried out in the fungus *Neurospora crassa*, the alga *Chlamydomonas reinhardtii,* the nematode *Caenorhabditis elegans,* and the plant *A. thaliana* to search for mutants defective in quelling, RNA interference, or PTGS. Analyses of these mutants led to the identification of host-encoded proteins involved in gene silencing and also revealed that a number of essential enzymes or factors are common to these processes. Some of the components identified serve as initiators, while others serve as effectors, amplifiers, and transmitters of the gene silencing process. In the years to come, many other components as well as their interrelations will be revealed.

RNase III family members are among the few nucleases that show specificity for dsRNAs and cleave them with 3′ overhangs of 2 to 3 nucleotides and 5′-phosphate and 3′-hydroxyl termini. Bernstein et al. identified an RNase III-like enzyme in *Drosophila* extract which was shown to have the ability to produce fragments of 22 nucleotides, similar to the size produced during RNAi. These authors showed that this enzyme is involved in the initiation of RNAi. Owing to its ability to digest dsRNA into uniformly sized small RNAs (siRNA), this enzyme was named Dicer (DCR). These nucleases are evolutionarily conserved in worms, flies, fungi, plants, and mammals. Dicer has four distinct domains: an amino-terminal helicase domain, dual RNase III motifs, a dsRNA binding domain, and a PAZ domain (a 110-amino-acid domain present in proteins like Piwi, Argo, and Zwille/Pinhead), which it shares with the RDE1/QDE2/Argonaute family of proteins that has been genetically linked to RNAi by independent studies. Cleavage by Dicer is thought to be catalyzed by its tandem RNase III domains. Some DCR proteins, including the one from *D. melanogaster,* contain an ATP-binding motif along with the DEAD box RNA helicase domain.

The predicted *C. elegans* Dicer homologue, K12H4.8, was referred as DCR1 because it was demonstrated to be the functional ortholog of the *Drosophila* Dicer protein. The 8,165-bp DCR1 protein has a domain structure similar to that of the *Drosophila* Dicer protein. *dcr1* mutants of *C. elegans* showed defects in RNAi of germ line-expressed genes but no effect on the RNAi response of somatic genes. These mutants were found to be sterile, suggesting the important role of this gene in germ line development apart from RNAi. CAF1 has been identified as a Dicer homologue in *A. thaliana,* but it is not involved in PTGS activity. The structure of CAF1 shows the presence of the four distinct domains that were identified in the *Drosophila* Dicer protein. Dicer homologues from many different sources have been identified; some recombinant Dicers have also been examined in vitro, and phylogenetic analysis of the known Dicer-like proteins indicates a common ancestry of these proteins.

Complete digestion by RNase III enzyme results in dsRNA fragments of 12 to 15 bp, half the size of siRNAs. The RNase III enzyme acts as a dimer and thus digests dsRNA with the help of two compound catalytic centers, whereas each monomer of the Dicer enzyme possesses two catalytic domains, with one of them deviating from the consensus catalytic sequences.

Recently, the crystal structure of the RNase III catalytic domain was solved, and this led to the model for generation of 23- to 28-mer diced siRNA products. In this model, the dimeric Dicer folds on the dsRNA substrate to produce four compound catalytic sites so that the two terminal sites having the maximum homology with the consensus RNase III catalytic sequence remain active, while the other two internal sites bearing partial homology lose functional significance. Thus, the diced products appear as the limit digests of the RNase III enzymes and are double the size of the normal 12- to 15-mer fragments. Such a model also predicts that certain changes in Dicer structure might modify the spacing between the two active terminal sites and thus generate siRNAs of variable sizes bearing species-specific imprints. Clearly, the crystal structure of Dicer is necessary to authenticate this model.

## Guide RNAs and RNA-induced Silencing Complex

Hammond et al. determined that the endogenous genes of *Drosophila* S2 cells could be targeted in a sequence-specific manner by transfection with dsRNA, and loss-of-function phenotypes were created in cultured *Drosophila* cells. The inability of cellular extracts treated with a $Ca^{2+}$-dependent nuclease (micrococcal nuclease, which can degrade both DNA and RNA) to degrade the cognate mRNAs and the absence of this effect with DNase I treatment showed that RNA was an essential component of the nuclease activity. The sequence-specific nuclease activity observed in the cellular extracts responsible for ablating target mRNAs was termed the RNA-induced silencing complex (RISC).

After partial purification of crude extracts through differential centrifugation and anion exchange chromatography, the nuclease cofractionated with a discrete ≈25-nucleotide RNA species. These results suggested that small RNAs were associated with sequence-specific nuclease and served as guides to target specific messages based upon sequence recognition. In another report, the multi-component RNAi nuclease was purified to homogeneity as a ribonucleoprotein complex of ≈500 kDa. One of the protein components of this complex was identified as a member of the Argonaute family of proteins and was termed Argonaute2 (AGO2). AGO2 is homologous to RDE1, a protein required for dsRNA-mediated gene silencing in *C. elegans*. AGO2 is a ≈130-kDa protein containing polyglutamine residues, PAZ, and PIWI domains characteristic of members of the Argonaute gene family. The Argonaute family members have been linked both to the gene-silencing phenomenon and to the control of development in diverse species. The first link between Argonaute protein and RNAi was shown by isolation of *rde1* mutants of *C. elegans* in a screen for RNAi-deficient mutants. Argonaute family members have been shown to be involved in RNAi in *Neurospora crassa* (QDE3) as well as in *A. thaliana* (AGO1).

Recently, two independent groups identified additional components of the RISC complex. Hammond and group showed the presence of two RNA binding proteins, the Vasa intronic gene and dFMR proteins, in the RISC complex isolated from *Drosophila* flies. Of these, dFMR is a homologue of the human fragile X mental retardation protein. In a parallel study, Siomi and group also isolated a novel ribonucleoprotein complex from the *Drosophila* lysate that contained dFMRI, AGO2, a *Drosophila* homologue of p68 RNA helicase (Dmp68), and two ribosomal proteins, L5 and L11, along with 5S rRNA. Both of these groups showed not only the presence of these components in the RISC complex, but also interactions among these proteins in vitro. Other components of RISC have not been clearly established yet.

## RNA and DNA Helicases

Aberrant RNA elimination surveillance seems to be common to most eukaryotic organisms. However, a diverse array of proteins specific for each organism seems to carry out such surveillance. Broadly, they fall in the biochemically similar group of RNA-DNA helicases. A mutant strain (*mut6*) of *C. reinhardtii* was isolated in which a gene required for silencing a transgene was disrupted. This RNAi-resistant mutant also showed an elevated transposition activity. The *mut6* gene was cloned and sequenced. The deduced MUT6 protein contains 1,431 amino acids and is a member of the DEAH box RNA helicase family. It also has a glycine-rich region that includes several RGG repeats, resembling an RGG box, a motif implicated in RNA binding and protein-protein interactions. MUT6 also has three putative nuclear localization signals and is predicted to be nuclear by PSORT analysis. MUT6 RNA helicase may be involved in degradation of misprocessed aberrant RNAs and thus could be a part of an RNAi-related surveillance system.

In *Neurospora crassa*, three classes of quelling-defective mutants (*qde1*, *qde2*, and *qde3*) have been isolated. The *qde3* gene has been cloned, and the sequence encodes a 1,955-amino-acid protein. The protein shows homology with several polypeptides belonging to the family of RecQ DNA helicases, which includes the human proteins for Bloom's syndrome and Werner's syndrome. In addition, QDE3 is believed to be involved in the activation step of gene silencing. The DNA helicase activity of QDE3 may function in the DNA-DNA interaction between introduced transgenes or with a putative endogenous gene required for gene-silencing activation by unwinding the double-stranded DNA. These interactions may induce changes in methylation or chromatin structure, producing an altered state that could result in aberrant RNA production. Thus, QDE3 protein may be more important for the transcriptional part of gene silencing, i.e., TGS.

When the RNAi sensitivity of several existing *C. elegans* mutants was examined, two mutant strains, *mut2* and *mut7*, that had previously shown elevated levels of transposon mobilization also showed resistance to RNAi. Ketting et al. identified a mutator gene, *mut7*, in *C. elegans* and characterized it at the molecular level. MUT7 was found to be homologous to proteins with 3'-5' exonuclease domains, such as Werner's syndrome protein and *E. coli* RNase D. It contained all the key catalytic residues for nuclease activity. A model was proposed in which MUT7 was speculated to play a role in repressing transposition by degrading the target mRNA with its exonuclease activity.

*smg* (suppressor of morphological effects on genitalia) mutants of *C. elegans*, defective in a process called nonsense-mediated decay, have been isolated. Seven *smg* genes which are involved in nonsense-mediated decay have been identified. Since this process also involves RNA degradation, the function of these genes, if any, in the RNAi process was examined. Animals mutant for a subset of these genes, *smg2*, *smg5*, and *smg6*, were initially silenced by dsRNA but later showed rapid recovery from the effects of RNAi, unlike the wild-type worms, which remained silenced. Thus, these genes might affect the persistence of RNA interference. On the other hand, *smg1*, *smg3*, and *smg4* mutant animals behaved like wild-type worms and did not recover from RNAi at all, indicating that these genes are not required for RNAi persistence. The *smg5* and *smg6* genes have not been cloned, but the *smg2* gene shows homology to *Saccharomyces cerevisiae upf1*, which encodes an ATPase with RNA-binding and helicase activities.

The SMG proteins could unwind dsRNA to provide a template for amplification activity. In this way, the three SMG proteins might facilitate amplification of the silencing signal and cause persistence of the silenced state. Alternatively, SMG proteins could increase the number of dsRNA molecules by promoting endonucleolytic cleavage of existing dsRNA molecules, which has been observed in *Drosophila* flies. No SMG2 homologues have been identified in plants or fungi. However, a search of the *A. thaliana* genome sequence database revealed a number of candidates with either helicase and RNase domains.

In a recent report, Tijsterman showed that unlike sense oligomers, single-stranded oligomers of antisense polarity could induce gene silencing in *C. elegans*. The antisense RNA-induced gene silencing was explained by proposing that RNA synthesis was primed on the mRNA by antisense RNA, resulting in dsRNAs, which acted as substrates for Dicer-dependent degradation. Antisense RNAs showed a requirement for the mutator/RNAi genes *mut7* and *mut14* but acted independently of the RNAi genes *rde1* and *rde4* of *C. elegans*. The *mut14* gene was cloned by genetic mapping and subsequent candidate gene approach. The MUT14 protein is a member of the family of putative RNA helicases that contain the signature DEAD box motif. These proteins are involved in diverse cellular functions. The helicase activity of MUT14 might thus act to permit de novo RNA synthesis on the target.

Dalmay et al. identified an *sde3* locus in *A. thaliana* plants which is required for the PTGS phenotype. They proposed that SDE3 protein might be involved in the production of dsRNA. SDE3 differs markedly from QDE3/MUT7 and has slight similarity to MUT6 in the helicase motif. Although it is highly similar to Upf1p and SMG2, it is unlikely that SDE3 is the functional homologue of Upf1p and SMG2 because it lacks important motifs. Notably, no SDE3 homologue was found in *C. elegans*, suggesting that SDE3-like proteins are regulators rather than essential cofactors of PTGS and are not used in *C. elegans*. This is further supported by the observation that *sde3* mutant plants exhibit only partial loss of PTGS. The closest homologue of SDE3 as identified by BlastP was a mouse protein encoded by *gb110*. These SDE3 homologues have RNA helicase motifs that are quite distinct from those of the DEAD, DEAH, and Ski2p types of RNA helicase. It has been speculated that SDE3 and SMG2 are multifunctional RNA helicases involved in PTGS.

## Translation Initiation Factor

Mutants of *C. elegans* showing resistance to dsRNA-mediated RNAi were selected by Tabara. They genetically mapped seven mutant strains that were placed in four complementation groups. One of the groups, *rde1*, consisted of three alleles. Gene *rde1* is a member of a large family which includes *Drosophila* homologues (*piwi* and *sting*) and *Arabidopsis* homologues (*argonaute* and *zwille*) and rabbit *eIF2C*. The full-length cDNA sequence for *rde1* was determined, and the deduced protein, consisting of 1,020 amino acids, was referred to as RDE1. The RDE1 protein is homologous to the product of the quelling deficiency (*qde2*) gene in *Neurospora crassa*. The initiation step of RNAi might be affected in the *rde1* mutant, as it completely lacks an interference response to several dsRNAs. It does not show any increase in transposon mobilization and or any effect on growth and development.

## RNA-dependent RNA Polymerase

The effects of both RNAi and PTGS are potent and systemic in nature. This has led to a proposed mechanism in which RNA-dependent RNA polymerases (RdRPs) play a role in both triggering and amplifying the silencing effect. Transgenic and virus-infected plants show an accumulation of aberrant transgenic and viral RNAs. The RdRP enzymes might recognize these aberrant RNAs as templates and synthesize antisense RNAs to form dsRNAs that are finally the targets for sequence-specific RNA degradation.

Genetic screens of *Neurospora crassa* (QDE1) and *A. thaliana* (SDE1/SGS2) led to the identification of proteins which are similar to tomato RdRP and are required for quelling and PTGS, respectively. This testifies to the importance of RdRP in gene silencing. Cogoni et al. cloned the *qde1* gene from *N. crassa*. It encodes a 158-kDa protein which lacks the typical signal peptide or a transmembrane domain, indicating its intracellular location. Dalmay found that the 113-kDa *Arabidopsis* RdRP is encoded by *sde1*. It is a plant homologue of QDE1 in *N. crassa* and EGO1 in *C. elegans*, which are required for quelling and RNAi, respectively. The SDE1 protein is required for transgene silencing but not for virus-induced PTGS, suggesting that SDE1 might be required to produce dsRNA, the initiator of PTGS.

The dsRNA produced as an intermediate in virus replication by virus-encoded RdRP might induce PTGS itself, and thus SDE1 may not be required for virus-induced PTGS. Plants with the *sde* mutation grow and develop normally, excluding a role for *sde* in development or basic cellular function. Two PTGS-controlling genes, *sgs2* and *sgs3*, were identified in *A. thaliana* by another group of workers. Later, it was found that *sgs2* and *sde1* are different descriptions of the same gene. On comparing the protein sequence of all the RdRPs, a conserved block was identified which seems to be crucial for RdRP function in PTGS and RNAi. *sgs3* mutants have the same molecular and phenotypic characteristics as *sgs2* mutants, but the SGS3 protein shows no significant similarity with any known putative proteins.

In *C. elegans*, EGO1, a protein required for RNAi, was found to be similar to tomato RdRP and the QDE1 protein of *Neurospora crassa*, as mentioned earlier. For a number of germ line-expressed genes, *ego1* mutants were resistant to RNA interference. The *ego1* transcript is found predominantly in the germ line. *ego1* is thus yet another example of a gene encoding an RdRP-related protein with an essential developmental function. RdRP is speculated to play a role in the amplification of the dsRNA signal, allowing its spread throughout the organism. The RdRP is also perhaps responsible for sustaining PTGS at the maintenance level even in the absence of the dsRNA that initiates the RNAi effect.

In spite of its omnipresence in different kinds of eukaryotic cells, RdRP homologues are not coded by either the *Drosophila* or human genome. Though the systemic characteristics of RNAi have not been revealed yet in either flies or humans, the amplification of siRNAs may be an essential step of RNAi even in these systems. Hence, it is important to know how these steps of RNAi are biochemically carried out in the absence of RdRP activity.

## Transmembrane Protein: Channel or Receptor

The systemic spread of gene silencing from one tissue to another has been well established in *C. elegans* and plants. To investigate the mechanism of systemic RNAi, Winston et al. constructed and used a special transgenic strain of *C. elegans* (HC57). They identified a systemic RNA interference-deficient (*sid*) locus required to transmit the effects of gene silencing between cells with green fluorescent protein (GFP) as a marker protein. Of the 106 *sid* mutants belonging to three complementation groups (*sid1*, *sid2*, and *sid3*), they isolated and characterized *sid1* mutants. The *sid1* mutants had no readily detectable mutant phenotype other than failure to show systemic RNAi. Interestingly, these mutants also failed to transmit the effect of RNAi to the progeny.

The SID1 polypeptide is predicted to be a 776 amino-acid membrane protein consisting of a signal peptide and 11 putative transmembrane domains. Based on the structure of SID1, it was suggested that it might act as a channel for the import or export of a systemic RNAi signal or might be necessary for endocytosis of the systemic RNAi signal, perhaps functioning as a receptor. No homologue of *sid1* was detected in *D. melanogaster*, which may be consistent with the apparent lack of systemic RNAi in the organism. However, the presence of SID homologues in humans and mice might hint at the systemic characteristics of RNAi in mammals.

## Genetic Mutations with Unknown Function

The three other complementation groups identified by Tabara et al. in *C. elegans* are *rde2* and *rde3*, with one allele each, and *rde4*, with two alleles. *rde4* mutants behaved like the *rde1* strain in not showing any increase in transposon mobilization and no effect on growth and development. The product of *rde2* remains to be identified. *mut2*, *rde2*, and *rde3* exhibited high-level transposition similar to *mut7*. This suggests a possible biological role of RNAi in transposon silencing.

Mello and colleagues have proposed that *rde1* and *rde4* respond to dsRNA by producing a secondary extragenic agent that is used by the downstream genes *rde2* and *mut7* to target specific mRNAs for PTGS. According to this view, *rde1* and *rde4* act as initiators of RNAi whereas *rde2* and *mut7* are effectors.

Table: Components of posttranscriptional gene silencing.

| Phenomenon | Organism | Mutation causing defective silencing | Gene function | Developmental defect |
|---|---|---|---|---|
| Posttranscriptional | Plant (*Arabidopsis thaliana*) | sgs2/sde1 | RdRP | None |
| gene silencing | | sgs3 | Unknown function | None |
| | | sde3 | RecQ helicase | |
| | | ago1 | Translation initiation factor | Pleiotropic effects on development & fertility |
| | | caf1 | RNA helicase & RNase III | |
| Quelling | Fungus (*Neurospora crassa*) | qde-1 | RdRP | None |
| | | qde-2 | Translation initiation factor | None |
| | | qde-3 | RecQ DNA helicase | |
| RNA interference | Worm (*Caenorhabditis elegans*) | ego-1 | RdRP | Gametogenetic defect & sterility |
| | | rde-1 | Translation initiation factor | None |
| | | rde-2, rde-3, rde-4, mut-2 | Unknown function | None |
| | | K12H4.8 (*dcr-1*) | Dicer homologue RNA helicase & RNase III | Sterility |
| | | mut-7 | Helicase & RNase D | None |
| | | mut-14 | DEAD box RNA helicase | |
| | | smg-2 | Upflp helicase | |
| | | smg-5 | Unknown function | |
| | | smg-6 | Unknown function | |
| | | sid-1 | Transmembrane protein | |
| | Alga (*Chlamydomonas reinhardtii*) | mut-6 | DEAH box RNA helicase | |

## Mechanism of RNA Interference

As the various pieces of the RNAi machinery are being discovered, the mechanism of RNAi is emerging more clearly. In the last few years, important insights have been gained in elucidating the mechanism of RNAi. A combination of results obtained from several in vivo and in vitro experiments have gelled into a two-step mechanistic model for RNAi/PTGS. The first step, referred to as the RNAi initiating step, involves binding of the RNA nucleases to a large dsRNA and its cleavage into discrete ≈21- to ≈25-nucleotide RNA fragments (siRNA). In the second step, these siRNAs join a multinuclease complex, RISC, which degrades the homologous single-stranded mRNAs. At present, little is known about the RNAi intermediates, RNA-protein complexes, and mechanisms of formation of different complexes during RNAi. In addition to several missing links in the process of RNAi, the molecular basis of its systemic spread is also largely unknown.

## Processing of dsRNA into siRNAs

Studies of PTGS in plants provided the first evidence that small RNA molecules are important intermediates of the RNAi process. Hamilton and Baulcombe, while studying transgene-induced PTGS in five tomato lines transformed with a tomato 1-aminocyclopropane-1-carboxyl oxidase (ACO), found accumulation of *aco* small RNAs of 25 nucleotides. More direct evidence about the generation of siRNAs in RNAi came from an in vitro cell-free system obtained from a *Drosophila* syncytial blastoderm embryo by Tuschl. These authors were able to reproduce many of the features of RNAi in this system. When dsRNAs radiolabeled within either the sense or the antisense strand were incubated with *Drosophila* lysate in a standard RNAi reaction, 21- to 23-nucleotide RNAs were generated with high efficiency. Single-stranded $^{32}$P-labeled RNA of either the sense or antisense strand was not efficiently converted to 21- to 23-nucleotide products. The formation of the 21- to 23-nucleotide RNAs did not require the presence of corresponding mRNAs.

The role of the small RNAs in RNAi was confirmed independently by Elbashir , who showed that synthetic 21- to 23-nucleotide RNAs, when added to cell-free systems, were able to guide efficient degradation of homologous mRNAs. To assess directly if the siRNAs were the true intermediates in an RNAi reaction, Zamore fractionated both the unprocessed dsRNAs and processed dsRNAs from the *Renilla luc* dsRNA-treated cell-free *Drosophila* system and showed that only the fractions containing native siRNAs were able to bring about the cognate RNA degradation and their ability to degrade RNA was lost when these fractions were treated at 95°C for 5 min. These in vivo and in vitro studies thus provided the evidence that siRNAs are the true intermediates of the RNAi reaction.

Together with the experiments to identify siRNAs as the key molecules for the RNAi effect, several investigators carried out the logical search for polypeptides that could generate such molecules. Based on the binding and cleavage properties of *E. coli* RNase III enzymes, Bass for the first time predicted the involvement RNase III-type endonucleases in the degradation of dsRNA to siRNAs. The RNase III enzyme makes staggered cuts in both strands of dsRNA, leaving a 3′ overhang of 2 nucleotides. The first evidence for the involvement of RNase III enzyme in RNAi was provided by T. Tuschl's group, who chemically analyzed the sequences of the 21- to 23-nucleotide RNAs generated by the processing of dsRNA in the *Drosophila* cell-free system. They showed the presence of 5′-phosphate, 3′-hydroxyl, and a 3′ 2-nucleotide overhang and no modification of the sugar-phosphate backbone in the processed 21- to 23-nucleotide RNAs.

Two groups recently identified candidate enzymes involved in degradation by scanning the genomes of *D. melanogaster* and *C. elegans* for genes encoding proteins with RNase III signatures. Bernstein showed that one of these identified genes, *dicer* in *Drosophila*, codes for the RNA processing enzyme that fragments dsRNA into 22-nucleotide fragments in vitro. An antiserum raised against Dicer could also immunoprecipitate a protein from the *Drosophila* extract or from S2 cell lysate, and these Dicer protein immunoprecipitates were able to produce RNAs of about 22 nucleotides from the dsRNA substrate. The direct correspondence in size of these RNAs with those generated from dsRNA by cell extract suggested a role of this protein in dsRNA degradation. The role of Dicer in RNAi was further confirmed by the fact that the introduction of Dicer dsRNA into *Drosophila* cells diminished the ability of the transfected cells to carry out RNAi in vitro. Similar experimental studies were carried out with *C. elegans* extract, and an ortholog of Dicer named DCR1 was identified.

A number of in vivo and in vitro experimental studies have shown that the production of 21- to 23-nucleotide RNAs from dsRNA requires ATP. The rate of 21- to 23-nucleotide RNA formation from corresponding dsRNAs has been shown to be six times slower in the *Drosophila* extract depleted for ATP by treatment with hexokinase and glucose. Bernstein and Ketting showed that the Dicer immunoprecipitates from *D. melanogaster* as well as S2 cell extracts and DCR1 immunoprecipitates from *C. elegans* extract required ATP for the production of 22-nucleotide RNAs. Recently, Nykanen reduced ATP levels in *Drosophila* extract by 5,000-fold with a sensitive ATP depletion strategy and showed considerable reduction in the rate of siRNA production in the *Drosophila* cell extract. These experiments suggest that ATP controls the rate of siRNA formation. However, it is still unclear whether ATP is absolutely rate limiting for the production of siRNAs from dsRNA.

The RNase activity and dsRNA binding of 218-kDa recombinant human Dicer have also been examined in vitro. The enzyme generated siRNA products from dsRNA quite efficiently in the presence of $Mg^{2+}$ and the absence of ATP. The RNase activity was sensitive to ionic interactions, whereas the dsRNA binding was quite effective in presence of high salt and did not require $Mg^{2+}$ at all. The dsRNA binding domain is located at the C terminus of Dicer, which is separable from the helicase and PAZ motifs. Human Dicer expressed in mammalian cells colocalized with calreticulin, a resident protein of the endoplasmic reticulum. In other systems, Dicer has also been found to complex with various other proteins. Hence, it is possible that the Dicer RNase activity functions as a complex of proteins in vivo.

## Amplification of siRNAs

One of the many intriguing features of RNA interference is the apparently catalytic nature of the phenomenon. A few molecules of dsRNA are sufficient to degrade a continuously transcribed target mRNA for a long period of time. Although the conversion of long dsRNA into many small siRNAs results in some degree of amplification, it is not sufficient to bring about such continuous mRNA degradation. Since mutations in genes encoding RNA-dependent RNA polymerase (RdRP) affect RNAi, it was proposed that this type owwf polymerase might replicate siRNAs as epigenetic agents, permitting their spread throughout plants and between generations in *C. elegans*. Recent studies by Lipardi) and Sijen provided convincing biochemical and genetic evidence that RdRP indeed plays a critical role in amplifying RNAi effects.

Lipardi while investigating the dsRNA-dependent degradation of target mRNA in a *Drosophila* embryo cell extract system, showed the generation of full-length cognate dsRNAs from labeled siRNAs at early time points. Both single-stranded RNAs (equivalent to target mRNA) and dsR-NAs served as templates for copying by RdRP. New full-length dsRNAs were formed rapidly and cleaved. They also showed a strict requirement for the 3'-hydroxyl group and 5'-phosphate group on siRNAs for primer extension in the RdRP-mediated reaction.

Sijen further revealed the role of RdRP activity in RNAi. In an RNAi reaction, they observed the formation of new siRNA species corresponding to target mRNAs but different from trigger dsR-NAs. They named these new siRNAs secondary siRNAs. With a primary trigger dsRNA specific for the *lacZ* region of the target mRNA that encoded a GFP-LacZ fusion protein, these authors demonstrated the degradation of a separate GFP mRNA target. This kind of RNAi induced by secondary siRNAs was named transitive RNAi. These authors demonstrated the requirement for the *rrf1* gene, a *C. elegans* gene with sequence homology to RdRP, in the generation of secondary siRNAs and transitive RNAi.

Amplification of siRNAs might occur at various stages of the RNAi reaction and has been documented in plants, *C. elegans*, *N. crassa*, and *Dictyostelium discoideum* but not in flies and mammals. Though the RdRP activity is present in *Drosophila* embryo extract, it is surprising that the fly genome does not code for RdRP. Additionally, numerous experiments also suggest that RdRP is not required for RNAi in *D. melanogaster*.

## Degradation of mRNA

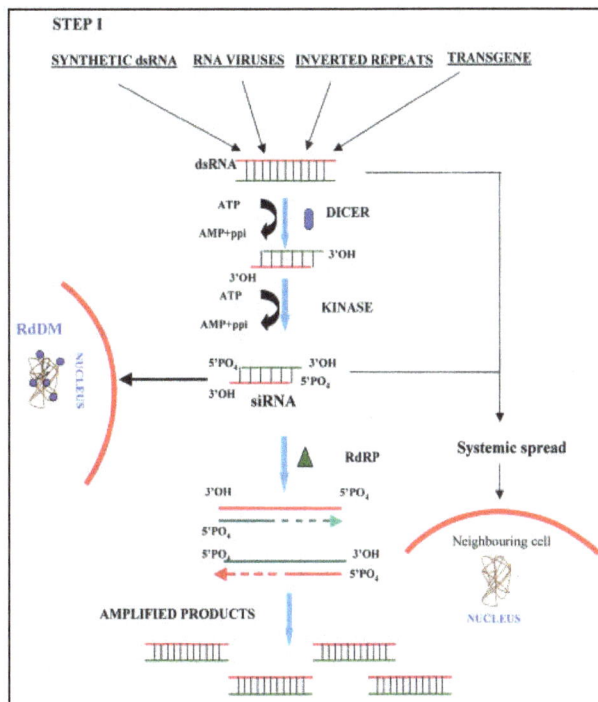

In the effector step of RNAi, the double-stranded siRNAs produced in the first step are believed to bind an RNAi-specific protein complex to form a RISC. This complex might undergo activation in the presence of ATP so that the antisense component of the unwound siRNA becomes exposed and allows the RISC to perform the downstream RNAi reaction. Zamore and colleagues demonstrated that a ≈250-kDa precursor RISC, found in *Drosophila* embryo extract, was converted into a ≈100-kDa complex upon being activated by ATP. This activated complex cleaved the substrate. The size and constitution of the precursor as well as the activated RISC might vary depending on the choice of system. The antisense siRNAs in the activated RISC pair with cognate mRNAs, and the complex cuts this mRNA approximately in the middle of the duplex region.

A few independent studies demonstrated the importance of the RISC complex in this part of RNAi reactions. The mRNA-cleaving RNA-protein complexes have also been referred to as siRNP (small interfering ribonucleoprotein particles). It is widely believed that this nuclease is probably different from Dicer, judging from the substrate requirements and the nature of the end products. Since the target cleavage site has been mapped to 11 or 12 nucleotides downstream of 5′ end of the guide siRNA, a conformational rearrangement or a change in the composition of an siRNP ahead of the cleavage of target mRNA is postulated. Finally, the cleaved mRNAs are perhaps degraded by exoribonucleases.

A part of cleaved fragments of mRNA at the end of step 2 might also be converted to the duplex forms by the RdRP-like activity. These forms might have siRNA-like functions and eventually enter the pool of the amplification reaction. Thus, it is likely that amplification of the RNAi reaction takes place at both step 1 and step 2 of RNAi. In another model, it has been proposed that siRNAs do not act as primers for the RdRP-like enzymes, but instead assemble along the length of the target RNA and are then ligated together by an RNA ligase to generate cRNA. The cRNA and target RNA hybrid would then be diced by the DCR protein. All these models were summarized by Schwarz. Most of the steps involved in the mechanism of RNAi have been illustrated schematically in figure.

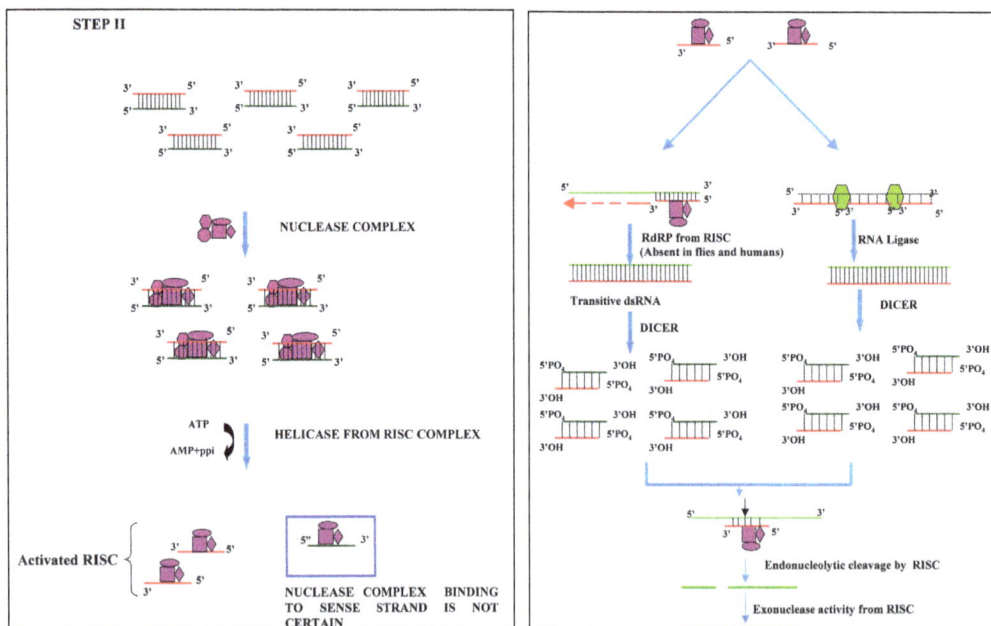

Two-step model for the mechanism of gene silencing induced by double-stranded RNA. In step I, dsRNA is cleaved by the Dicer enzyme to produce siRNAs. A putative kinase seems to maintain 5′ phosphorylation at this step. The siRNAs have also been proposed to be responsible for nuclear DNA methylation and systemic spread of silencing. Amplification might occur due to the presence of RdRP. In step II, the siRNAs generated in step I bind to the nuclease complex (RISC). A helicase present in the complex might activate RISC by unwinding the siRNAs. The antisense component of siRNA in the RISC guides the complex towards the cognate mRNA, resulting in endonucleolytic cleavage of the mRNA. RdDM, RNA-dependent DNA methylation.

## RNA Silencing for Genome Integrity and Defense

Considerable evidence indicates that PTGS has evolved as a protective mechanism against parasitic DNA sequences such as transposons and the RNA sequences of plant viruses. DNA methylation and transcriptional gene silencing (TGS) are mainly responsible for keeping the transposition frequency at a minimum. However, PTGS also provides additional protection against the genomic instability caused by transposons. Mutations in the *C. elegans mut-7* gene increase the transposition frequency in the germ line and downregulate RNAi as well, implicating RNAi in the control of transposons. Recently, Djikeng cloned and sequenced the siRNA products of an RNA interference event occurring in *Trypanosoma brucei*. By sequencing over 1,300 siRNA-like fragments, they observed abundant 24- to 26-nucleotide fragments homologous to the ubiquitous retrotransposon INGI and the site-specific retroposon SLACS. Thus, they convincingly demonstrated that RNAi is involved in silencing the retroposon transcript.

In plants, PTGS has been widely linked with RNA virus resistance mechanisms. Plant RNA viruses are, in fact, both inducers and targets for PTGS and gene-silencing-defective mutants of plants show increased sensitivity to viral infections. The direct role of dsRNA in inhibiting viral infection has recently been demonstrated by Tenllado and Diaz-Ruiz. They showed that dsRNAs derived from viral replicase sequences could interfere with virus infection in a sequence-specific manner by directly delivering the dsRNAs to leaf cells either by mechanical coinoculation with the virus or via an *Agrobacterium*-mediated transient-expression approach. Successful interference with the infection of plants by representative viruses belonging to the tobamovirus, potyvirus, and alfamovirus genera has been demonstrated. These results support the view that a dsRNA intermediate in virus replication acts as an efficient initiator of PTGS in natural virus infections.

The clinching support for the notion that PTGS has evolved as an antiviral mechanism has come from reports that plant viruses encode proteins that are suppressors of PTGS. These suppressors have evolved to save the viral RNA genomes from the PTGS degradative machinery of host plants. Different types of viral suppressors have been identified through the use of a variety of silencing suppression assays. Suppressors HC-PRO, P1, and AC2 are one type (encoded by potyviruses, rice yellow mottle sobemovirus, and geminiviruses of subgroup III, respectively) that is able to activate GFP expression in all tissues of previously silenced GFP-expressing plants. HC-PRO reduces target mRNA degradation and is thus responsible for reduced accumulation of siRNAs. The second type of suppressors includes movement proteins, i.e., p25 of potato virus X, which are involved in curbing the systemic aspect of transgene-induced RNA silencing. The third type includes cytomegalovirus 2b protein, which is involved in systemic signal-mediated RNA silencing. The cytomegalovirus 2b protein is nucleus localized and also inhibits salicylic acid-mediated virus resistance. Other types of viral suppressors with undefined biochemical activities are also known. These findings not only provide the strongest support that PTGS functions as a natural, antiviral defense mechanism, but also offer valuable tools for dissecting the biochemical pathways of PTGS.

The PTGS degradative machinery can both detect and inactivate repetitive DNA sequences, suggesting it controls the expansion of repetitive elements, including endogenous genes. Although

RNAi occurs in mammals and mammalian cell cultures, its role in animal virus protection is not clear. In mammals, dsRNA induces RNAi as well as interferon-mediated nonspecific RNA degradation and other nonspecific responses leading to blockage in protein synthesis and cell death. Thus, mammals seem to have evolved multiple mechanisms to detect and target dsRNA and to fight viruses. These various mechanisms may have different specificities or can function in distinct tissues or during development.

## Applications of RNAi

Besides being an area of intense, upfront basic research, the RNAi process holds the key to future technological applications. Genome sequencing projects generate a wealth of information. However, the ultimate goal of such projects is to accelerate the identification of the biological function of genes. The functions of genes can be analyzed with an appropriate assay, by examining the phenotype of organisms that contain mutations in the gene, or on the basis of knowledge gained from the study of related genes in other organisms. However, a significant fraction of genes identified by the sequencing projects are new and cannot be rapidly assigned functions by these conventional methods.

RNAi technology is proving to be useful to analyze quickly the functions of a number of genes in a wide variety of organisms. RNAi has been adapted with high-throughput screening formats in *C. elegans*, for which the recombination-based gene knockout technique has not been established. Chromosomes I and III of *C. elegans* have been screened by RNAi to identify the genes involved in cell division and embryonic development. Recently, a large-scale functional analysis of ≈19,427 predicted genes of *C. elegans* was carried out with RNA interference. This study identified mutant phenotypes for 1,722 genes. Similarly, in *D. melanogaster*, RNAi technology has been successfully applied to identify genes with essential roles in biochemical signaling cascades, embryonic development, and other basic cellular process. In plants, gene knockdown-related functional studies are being carried out efficiently when transgenes are present in the form of hairpin (or RNAi) constructs. Plant endotoxins could also be removed if the toxin biosynthesis genes are targeted with the RNAi constructs. Recently, the theobromine synthase of the coffee plant was knocked down with the hairpin construct of the transgene, leading to the production of decaffeinated coffee plants. Virus-induced gene silencing has also been proven to be a successful approach for plant genetics.

Given the fact that RNAi is easy to apply, whole-genome screens by RNAi may become a common method of choice in the near future. RNAi may facilitate drug screening and development by identifying genes that can confer drug resistance or genes whose mutant phenotypes are ameliorated by drug treatment, providing information about the modes of action of novel compounds. Although RNAi is unlikely to replace the existing knockout technology, it may have a tremendous impact for those organisms that are not amenable to the knockout strategy. It may also be a method of choice to study the simultaneous functions of a number of analogous genes in organisms in which redundancy exists with respect to a particular function, because many of these genes can be silenced simultaneously.

Given the gene-specific features of RNAi, it is conceivable that this method will play an important role in therapeutic applications. Since siRNAs direct cellular RNAi biology, these are potential therapeutic reagents because of their power to downregulate the expression pattern of mutant genes in diseased cells. However, central to this hypothesis is the assumption that the effect of exogenous siRNA applications will remain gene specific and show no nonspecific side effects relating to mismatched off-target hybridization, protein binding to nucleic acids, etc. Though it was demonstrated that mismatches of more than even one nucleotide within the 19- to 20-mer siRNAs effectively disrupted proper degradation of the target mRNA, the gene specificity of siRNAs needs to be confirmed on a genome-wide scale.

Recently, Chi reported that the GFP siRNA-induced gene silencing of transient or stably expressed GFP mRNA was highly specific in the human embryonic kidney (HEK) 293 cell background. The specific silencing did not produce secondary changes in global gene expression, as detected by the DNA microarray experiment. They also failed to detect the presence of transitive RNAi in experimentally engineered human cell lines. In their own experiments, Semizarov reached a similar conclusion while using siRNAs corresponding to *akt1*, *rb1*, and *plk1* in the human non-small cell lung carcinoma cell line H1299. These experiments prove that siRNAs could be used as highly specific tools for targeted gene knockdown and can be used in high-throughput approaches and drug target validation. This exquisite sequence-specific effect of siRNAs has also been exploited in silencing the mutant allele of the diseased gene while not affecting the wild-type allele of the healthy version of the same gene.

siRNAs have been shown to inhibit infection by human immunodeficiency virus, poliovirus, and hepatitis C virus in cultured cell lines. Bitko and Barik successfully used siRNAs to silence genes expressed from respiratory syncytial virus, an RNA virus that causes severe respiratory disease in neonates and infants. siRNA treatment has also been shown to reduce the expression of the BCR-ABL oncoprotein in leukemia and lymphoma cell lines, leading to apoptosis in these cells. With respect to future medical applications, siRNA-based therapy seems to have a great potential to combat carcinomas, myeloma, and cancer caused by overexpression of an oncoprotein or generation of an oncoprotein by chromosomal translocation and point mutations.

Recently, the therapeutic potential of the siRNA technique has been demonstrated in vivo in mouse models. McCaffrey and Song demonstrated effective targeting of a sequence from hepatitis C virus and the *fas* gene by RNA interference in mouse liver. An epiallelic series of p53 hypomorphs created by RNAi have been shown to produce distinct tumor phenotypes in mice in vivo, suggesting that RNAi can stably suppress gene expression. Song has shown that treatment with *fas* siRNA abrogated hepatocyte necrosis and inflammatory infiltration and protected mice from liver fibrosis and fulminant hepatitis. Rubinson showed highly specific, stable, and functional silencing of gene expression in transgenic mice with the lentivirus system for the delivery of siRNAs.

Although the delivery of siRNAs to a proper site remains problematic for gene therapy, chemical modifications of siRNAs such as changing the lipophilicity of the molecules or the methods previously developed for the application of antisense oligonucleotides or nuclease-resistant ribozymes might help the entry and stability of siRNAs within the transfected cells or tissues. The absence of specific micro-RNAs has been demonstrated in carcinoma cells, implying that cancer development could be arrested by introduction of the missing micro-RNAs. The micro-RNAs could be supplied in the form of siRNAs, since the function of micro-RNAs can be mimicked by the exogenous siRNA. However, independent of its biomedical applications, RNAi appears to be a forthcoming method for functional genomics.

# CRISPR

CRISPR is an acronym for Clustered Regularly Interspaced Short Palindromic Repeat. This name refers to the unique organization of short, partially palindromic repeated DNA sequences found in the genomes of bacteria and other microorganisms. While seemingly innocuous, CRISPR sequences are a crucial component of the immune systems of these simple life forms. The immune system is responsible for protecting an organism's health and well-being. Just like us, bacterial cells can be invaded by viruses, which are small, infectious agents. If a viral infection threatens a bacterial cell, the CRISPR immune system can thwart the attack by destroying the genome of the invading virus. The genome of the virus includes genetic material that is necessary for the virus to continue replicating. Thus, by destroying the viral genome, the CRISPR immune system protects bacteria from ongoing viral infection.

## Working of CRISPR

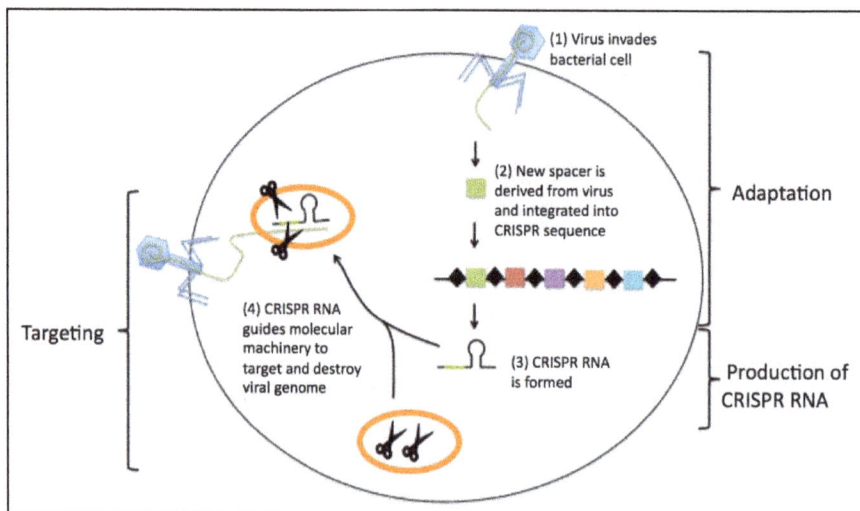

Figure: The steps of CRISPR-mediated immunity.

CRISPRs are regions in the bacterial genome that help defend against invading viruses. These regions are composed of short DNA repeats (black diamonds) and spacers (colored boxes). When a previously unseen virus infects a bacterium, a new spacer derived from the virus is incorporated amongst existing spacers. The CRISPR sequence is transcribed and processed to generate short CRISPR RNA molecules. The CRISPR RNA associates with and guides bacterial molecular machinery to a matching target sequence in the invading virus. The molecular machinery cuts up and destroys the invading viral genome.

Interspersed between the short DNA repeats of bacterial CRISPRs are similarly short variable sequences called spacers. These spacers are derived from DNA of viruses that have previously attacked the host bacterium. Hence, spacers serve as a 'genetic memory' of previous infections. If another infection by the same virus should occur, the CRISPR defense system will cut up any viral DNA sequence matching the spacer sequence and thus protect the bacterium from viral attack. If a previously unseen virus attacks, a new spacer is made and added to the chain of spacers and repeats.

The CRISPR immune system works to protect bacteria from repeated viral attack via three basic steps:

Step 1) Adaptation – DNA from an invading virus is processed into short segments that are inserted into the CRISPR sequence as new spacers.

Step 2) Production of CRISPR RNA – CRISPR repeats and spacers in the bacterial DNA undergo transcription, the process of copying DNA into RNA (ribonucleic acid). Unlike the double-chain helix structure of DNA, the resulting RNA is a single-chain molecule. This RNA chain is cut into short pieces called CRISPR RNAs.

Step 3) Targeting – CRISPR RNAs guide bacterial molecular machinery to destroy the viral material. Because CRISPR RNA sequences are copied from the viral DNA sequences acquired during adaptation, they are exact matches to the viral genome and thus serve as excellent guides.

The specificity of CRISPR-based immunity in recognizing and destroying invading viruses is not just useful for bacteria. Creative applications of this primitive yet elegant defense system have emerged in disciplines as diverse as industry, basic research, and medicine.

## Some Applications of the CRISPR System

### In Industry

The inherent functions of the CRISPR system are advantageous for industrial processes that utilize bacterial cultures. CRISPR-based immunity can be employed to make these cultures more resistant to viral attack, which would otherwise impede productivity. In fact, the original discovery of CRISPR immunity came from researchers at Danisco, a company in the food production industry. Danisco scientists were studying a bacterium called *Streptococcus thermophilus*, which is used to make yogurts and cheeses. Certain viruses can infect this bacterium and damage the quality or quantity of the food. It was discovered that CRISPR sequences equipped *S. thermophilus* with immunity against such viral attack. Expanding beyond *S. thermophilus* to other useful bacteria, manufacturers can apply the same principles to improve culture sustainability and lifespan.

### In the Lab

Beyond applications encompassing bacterial immune defenses, scientists have learned how to harness CRISPR technology in the lab to make precise changes in the genes of organisms as diverse as fruit flies, fish, mice, plants and even human cells. Genes are defined by their specific sequences, which provide instructions on how to build and maintain an organism's cells. A change in the sequence of even one gene can significantly affect the biology of the cell and in turn may affect the health of an organism. CRISPR techniques allow scientists to modify specific genes while sparing all others, thus clarifying the association between a given gene and its consequence to the organism.

Rather than relying on bacteria to generate CRISPR RNAs, scientists first design and synthesize short RNA molecules that match a specific DNA sequence—for example, in a human cell. Then, like in the targeting step of the bacterial system, this 'guide RNA' shuttles molecular machinery to the intended DNA target. Once localized to the DNA region of interest, the molecular machinery can silence a gene or even change the sequence of a gene. This type of gene editing can be likened to editing a sentence with a word processor to delete words or correct spelling mistakes. One important application of such technology is to facilitate making animal models with precise genetic changes to study the progress and treatment of human diseases.

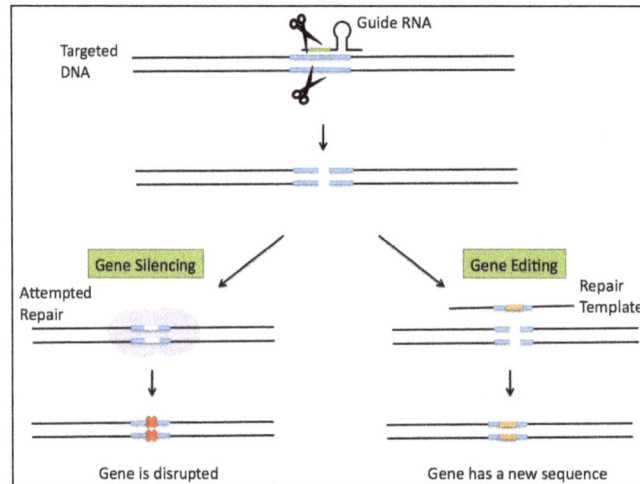

Figure: Gene silencing and editing with CRISPR.

Guide RNA designed to match the DNA region of interest directs molecular machinery to cut both strands of the targeted DNA. During gene silencing, the cell attempts to repair the broken DNA, but often does so with errors that disrupts the gene—effectively silencing it. For gene editing, a repair template with a specified change in sequence is added to the cell and incorporated into the DNA during the repair process. The targeted DNA is now altered to carry this new sequence.

## In Medicine

With early successes in the lab, many are looking toward medical applications of CRISPR technology. One application is for the treatment of genetic diseases. The first evidence that CRISPR can be used to correct a mutant gene and reverse disease symptoms in a living animal was published earlier this year. By replacing the mutant form of a gene with its correct sequence in adult mice, researchers demonstrated a cure for a rare liver disorder that could be achieved with a single treatment. In addition to treating heritable diseases, CRISPR can be used in the realm of infectious diseases, possibly providing a way to make more specific antibiotics that target only disease-causing bacterial strains while sparing beneficial bacteria.

## Future of CRISPR

Of course, any new technology takes some time to understand and perfect. It will be important to verify that a particular guide RNA is specific for its target gene, so that the CRISPR system does not mistakenly attack other genes. It will also be important to find a way to deliver CRISPR therapies into the body before they can become widely used in medicine. Although a lot remains to be discovered, there is no doubt that CRISPR has become a valuable tool in research. In fact, there is enough excitement in the field to warrant the launch of several Biotech start-ups that hope to use CRISPR-inspired technology to treat human diseases.

# Permissions

# Index